The Minneapolis General Math Project

A REASONABLY CLOSE ENCOUNTER

Estimation Skills

The Minneapolis General Math Project was developed by
the Minneapolis Public Schools, Minneapolis, Minnesota.

Project Co-Directors: Pamela Katzman
Jackie McClees

Production and Artwork: Ingrid Holley

Project Team: Cheri Carlson
Paul Dillenberger
Al Hoogheem
Mike Hughes
Mary Indelicato
Bill Robbins
Marion Thorne
Les Twedell

Contributors: Anne Bartel
Carol Borne
Vicki DeVoss
Ron Fish
John Hendrickson
Barb Kitlinski
John Maus
Debii (Winberg) Nelson

**Mathematics Consultant,
Minneapolis Public Schools:** Ross Taylor

The Project Team wishes to thank those teachers and students
in the Minneapolis Public School System who field-tested and
evaluated these materials as they were being developed.

ISBN 0-86651-117-2

Order Number DS01391

cdefghi-MA-9543210

CONTENTS

THE MINNEAPOLIS GENERAL MATH PROJECT

What is the Minneapolis General Math Project?

The Minneapolis General Math project consists of 8 different books containing a total of 61 teaching units. The teaching units were developed to supplement the teaching of basic mathematical skills as identified by the National Council of Supervisors of Mathematics (NCSM). The lessons in each unit reinforce the use of basic math skills while exploring topics which interest secondary school students.

Each of the books in the Minneapolis General Math Project corresponds to one or more basic skill areas.

BOOK	BASIC SKILLS
In a Word . . . Checking Out Patterns	Problem Solving
The Whole of It Part and Parcel Watch Your Wallet	Applying Mathematics to Everyday Situations Appropriate Computational Skills
A Reasonably Close Encounter	Alertness to Reasonableness of Results Estimation and Approximation
The Size of It	Geometry Measurement
Charting Your Course	Reading, Interpreting, and Constructing Tables, Charts, and Graphs

Two basic skill areas identified by the NCSM, Using Mathematics to Predict and Computer Literacy, are *not* reinforced in the books. There are other materials available that thoroughly cover these skills.

Who can use the Minneapolis General Math Project?

The lessons in the Minneapolis General Math Project were designed
for ninth-grade general mathematics courses. They are appropriate
for use in any secondary school situation where reinforcing the use of
basic math skills is desired. The lessons also can be used with adults
who are earning high school equivalency certificates.

What do the books contain?

Each book contains from 4 to 11 units. A unit consists of blackline
master lesson pages which focus on a particular topic of interest such
as earthquakes, maps, tipping, or palindromes. There is a teacher's
page describing the contents of the unit and giving suggestions for
its use. Some units include a test. An answer key (which includes
answers for test items) is provided for each unit.

How can the books be used?

The units of the Minneapolis General Math Project can be used inde-
pendently of one another for individual, group, or full-class activities,
depending on your classroom organization and needs. You may select
the units you wish to use and teach them in different orders. In some
cases, a particular order of presentation is recommended.

The books are designed so the pages can be easily removed and
reproduced by Thermofax, Xerox, or a similar process. You can make
either transparencies for overhead projection or pages for student use.

The lessons within a unit are sequential. Most teachers copy all the
lesson pages for a given unit at once and provide one set for each
student to be used as consumable workbooks. You can make the units
non-consumable by creating a blank answer sheet for your students.
However, any puzzles should be reproduced for individual student use.

CONTENTS OF THE PROJECT

The following chart shows the books and units for the complete Minneapolis General Math Project. All units reinforce the use of whole numbers. Some units also reinforce the use of fractions, decimals, percent, or signed numbers.

	WHOLE NOS.	FRACTIONS	DECIMALS	PERCENT	SIGNED NOS.
IN A WORD . . .					
Operation: Word Problem	✔		✔		
De-Vine Word Problems	✔		✔		
Read & Reason	✔		✔		
Calculators with a Twist	✔		✔		
CHECKING OUT PATTERNS					
Simple Sequences	✔				
Sequences	✔	✔			
Fibonacci Sequence	✔				
Fibonacci Fractions	✔	✔	✔		
Picture This	✔				
We The Jury	✔	✔			
Easy Does It	✔				
Fanciful Formulas	✔	✔			
THE WHOLE OF IT					
Mom, Dad, Bob & Lil	✔				
The Key to Locks	✔				
Telephone Trivia	✔				
License Plates	✔				
MPG	✔				
Calorie Countdown	✔				
Calories Count Up	✔				
Earthquakes	✔				
PART AND PARCEL					
Buttonholes	✔	✔			
A Fishy Tale	✔	✔			
Stock Up	✔	✔			
Pizza for Breakfast	✔	✔			
The Fraction Link	✔	✔			
Right Off the Bat	✔	✔	✔		
Hannah, Anna, Otto & Ada	✔		✔		
Carat with a "C"	✔		✔		
Karat with a "K"	✔	✔		✔	
Bionic Betty	✔			✔	
Tipping	✔		✔	✔	

	WHOLE NOS.	FRACTIONS	DECIMALS	PERCENT	SIGNED NOS.
WATCH YOUR WALLET					
You Can Bank On It	✔				
Money Matters	✔				
Sales Tax	✔				
Wise Buys	✔				
A REASONABLY CLOSE ENCOUNTER					
About "About"	✔				
Making Some Headway	✔				
Making an Adjustment	✔				
Just Around the Corner	✔		✔		
Be Reasonable	✔		✔		
Divide and Conquer	✔				
Everyday Estimation	✔		✔		
Making You Percentable	✔	✔	✔	✔	
Fraction Action	✔	✔	✔		
The Money Round-Up	✔		✔		
THE SIZE OF IT					
Getting Around	✔				
Horsing Around	✔		✔		
Cover Up	✔				
State of Bedlam	✔		✔		
Maximum's Area	✔				
Perimeter Is Your Area	✔		✔		
Pizza Pi	✔		✔		
Pan Plan	✔		✔		
Temperature	✔				
The Weather Report	✔				✔
Is It On the Level?	✔				✔
CHARTING YOUR COURSE					
Come to Grips with Graphs	✔	✔	✔	✔	
Road Distances	✔				
Travel Tips	✔	✔	✔		
Day Finder	✔				
Is It Your Day?	✔				

CONTENTS OF
A REASONABLY CLOSE ENCOUNTER

gmp general math project

minneapolis public schools

UNIT: ABOUT "ABOUT"

DESCRIPTION: This unit is an introduction to the concept of estimation.

SKILLS TO BE REINFORCED: estimating and problem solving skills

PREREQUISITE SKILLS: none

LENGTH: 4 pages;   less than 1 day

TEACHER NOTES: This is a mere beginning of work on estimation. The following sequence of units is recommended:

        MAKING SOME HEADWAY

        MAKING AN ADJUSTMENT

        'JUST AROUND THE CORNER

        BE REASONABLE

        DIVIDE AND CONQUER

        EVERYDAY ESTIMATION

        MAKING YOU PERCENTABLE!

        FRACTION ACTION

        THE MONEY ROUND-UP

No methods are presented in this unit for estimating. However, students do have to estimate. To introduce the idea of estimating, you may want to discuss how the students arrived at their estimations for the problems on pages 3 and 4.

NO TEST

ABOUT "ABOUT"!

Think about the word "ABOUT" for about a minute.   Do you ever think or say these kinds of things?

> It will take me <u>about</u> 10 minutes . . .
>
> It will cost <u>about</u> six dollars . . .
>
> I'll need <u>about</u> 6 rolls of wallpaper . . .
>
> It's <u>about</u> 3 miles from here . . .
>
> It weighs <u>about</u> 10 pounds more . . .
>
> Our car gets <u>about</u> 20 miles to the gallon . . .

Now make a good guess about how many times you use the word "ABOUT" every day! ___________________
(1)

and think about it !

When you leave the house to catch the school bus or to walk to school, does it take you exactly the same length of time every day or <u>about</u> the same time? _______________ (2)

The sun is 93,000,000 miles from Earth. Is that an exact distance or does it tell <u>about</u> how far we are from the sun? _______________ (3)

When you go to the grocery store, do you usually take the exact amount of money, or <u>about</u> what you think the right amount will be? _______________ (4)

When you pour a half a glass of milk, do you pour exactly a half glass or <u>about</u> a half glass?

_______________ (5)

Now list 3 more examples of situations in which <u>about</u> is close enough!

1. ____
____ (6)

2. ____
____ (7)

3. ____
____ (8)

    About "About" 2

Now do some easy estimates. Just use your head, make a good guess, and circle the best answer.

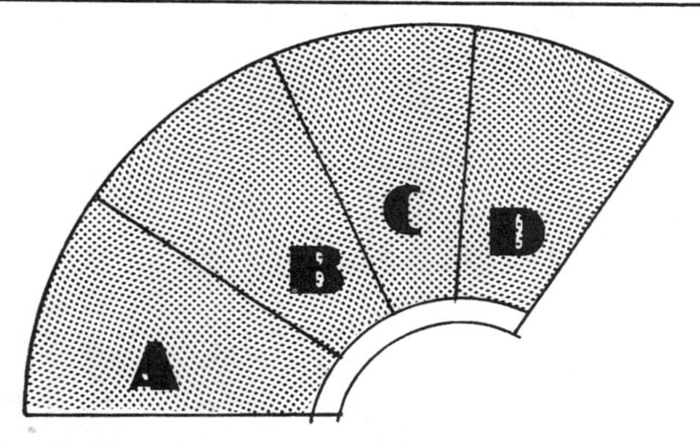

At a rock concert, section D was about half full and sections A, B, and C were almost sold out.

If each section of the stadium has 3000 seats, <u>about</u> how many people attended the rock concert?

8000   10,500   12,000

(9)

Before you can go out on Friday evening, you have to do the dishes, sweep the kitchen floor and change your clothes. <u>About</u> how long do you think it would take you to do these things?

10 minutes   45 minutes   2 hours

(10)

You are going to the grocery store to buy a carton of milk, a loaf of bread, a package of lunch meat and a head of lettuce. <u>About</u> how much money would you need to take?

$1 $5 $10

(11)

<u>About</u> how much of this figure is shaded?

1/8   1/2   5/6

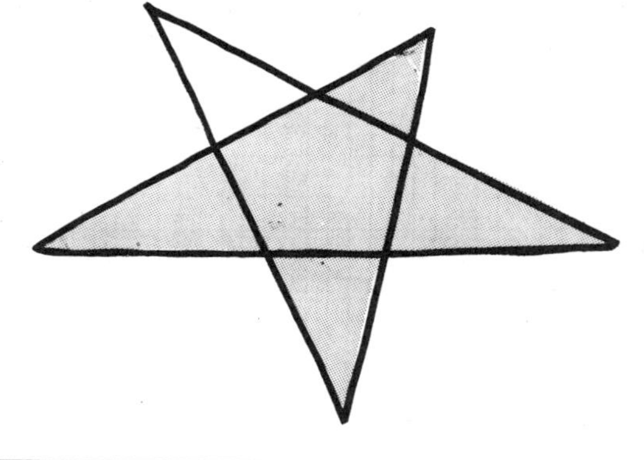

(12)

You work 20 hours a week. Your salary is $3.15 per hour. Your paycheck for <u>two</u> weeks should be <u>about</u> . . .

$600.00 $120.00 $60.00

$12.00 $1.20

(13)

One wall of a library is pictured here. One shelf of one section holds about 20 books.

Estimate <u>about</u> how many books are in one section. ____________ (14)

Estimate <u>about</u> how many books are in all five sections. ____________ (15)

As you learn to estimate you will learn:

- to tell about how many of something there are without counting each one

- to make good guesses (called "educated guesses") about things we usually measure

- to adjust numbers and to quickly estimate in your head the answers to problems

- to use your judgement about how close to the exact answer an estimated answer should be

UNIT _ABOUT "ABOUT"_   ANSWER KEY

1. _answers will vary_
2. _about_
3. _about_
4. _about_
5. _about_
6. _answers will vary_
7. _answers will vary_
8. _answers will vary_
9. _10,500_
10. _45 minutes_
11. _$5_
12. _5/6_
13. _$120.00_
14. _120_
15. _500-600_
16. _______________
17. _______________
18. _______________
19. _______________
20. _______________
21. _______________
22. _______________
23. _______________
24. _______________
25. _______________

26. _______________
27. _______________
28. _______________
29. _______________
30. _______________
31. _______________
32. _______________
33. _______________
34. _______________
35. _______________
36. _______________
37. _______________
38. _______________
39. _______________
40. _______________
41. _______________
42. _______________
43. _______________
44. _______________
45. _______________
46. _______________
47. _______________
48. _______________
49. _______________
50. _______________

gmp general math project

minneapolis public schools

UNIT: MAKING SOME HEADWAY

DESCRIPTION: When you estimate, you need to compute quickly in your head. This unit gives students practice in doing that.

SKILLS TO BE REINFORCED: addition, subtraction, multiplication, and division of whole numbers

PREREQUISITE SKILLS: basic addition, subtraction, multiplication and division facts

LENGTH: 8 pages;   3 - 4 days

TEACHER NOTES: This should be thought of as a way to compute answers after the numbers have been adjusted.

'Adjusting' numbers includes Truncation, Left-End Rounding and Rounding. These methods of adjusting numbers are taught in MAKING AN ADJUSTMENT and 'JUST AROUND THE CORNER.

NO TEST

MAKING SOME HEADWAY

2000 + 11,000 + 8000 + 3000 = ?

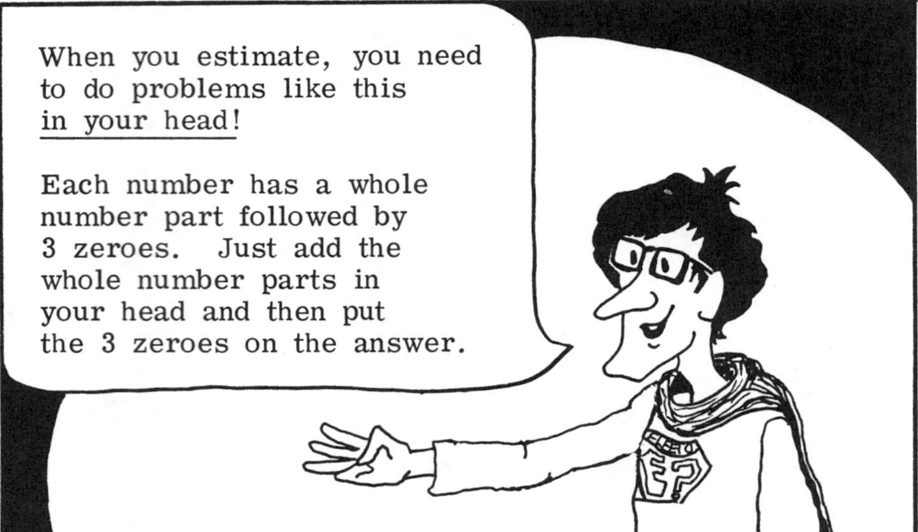

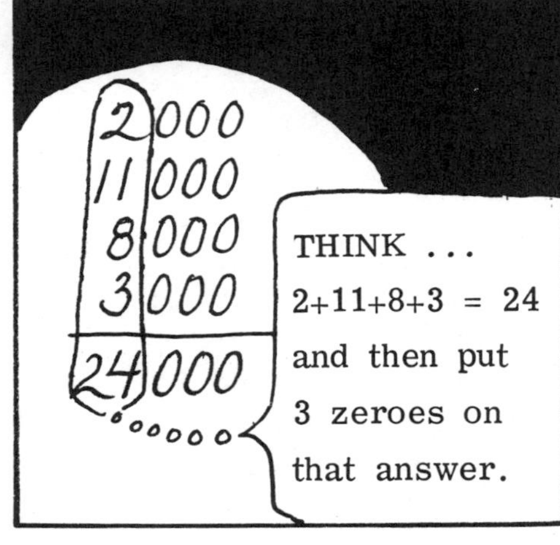

When you add numbers which contain the same number of 0's, follow these steps in your head:

① Add the whole number parts in your head.

$$4+9+14+2=29$$

$$\begin{array}{r} 4000 \\ 9000 \\ 14000 \\ +\ 2000 \\ \hline 29000 \end{array}$$

② Then put the same number of 0's on the answer.

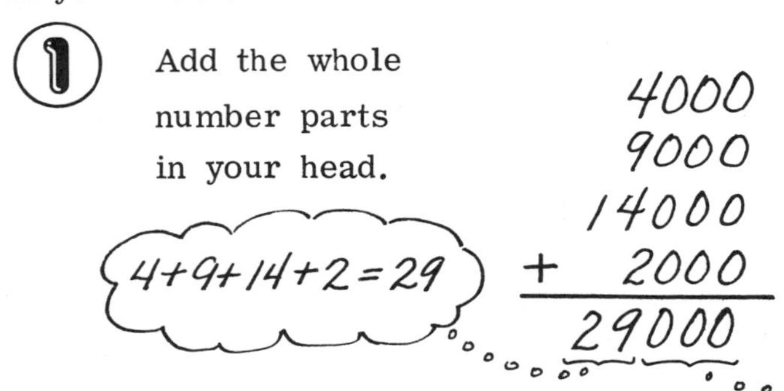

1100 + 600 + 800 =  **?**

What are the whole number parts of the numbers?

____ , ____ , and ____
(1)   (2)   (3)

Does each number contain the same number of 0's? ____
How many? ____
(4)
(5)

1 Add the whole number parts in your head.

$$1100$$
$$600$$
$$+ \ 800$$

2 Then put the same number of 0's on the answer.

5000 + 13,000 + 2000 + 1000 = ?

What are the whole number parts of the numbers?

____ , ____ , ____ , and ____
(8)   (9)   (10)   (11)

How many 0's does each number have? ____
(12)

1 Add the whole number parts in your head.

$$5000$$
$$13000$$
$$2000$$
$$+ \ 1000$$

____ ____
(13) (14)

2 Then put the same number of 0's on the answer.

Making Some Headway 2

Add in your head!

$22,000 + 5000 + 8000 =$ _________ (15)

$30 + 50 + 70 =$ _________ (16)

$1500 + 200 + 500 =$ _________ (17)

$700 + 100 + 700 =$ _________ (18)

$6000 + 11,000 + 4000 =$ _________ (19)

Now try these...

★ $2000 + 900 =$ ❓

Think of both numbers as having 2 zeroes.

2000 + 900

Then the whole number parts are 20 and 9.

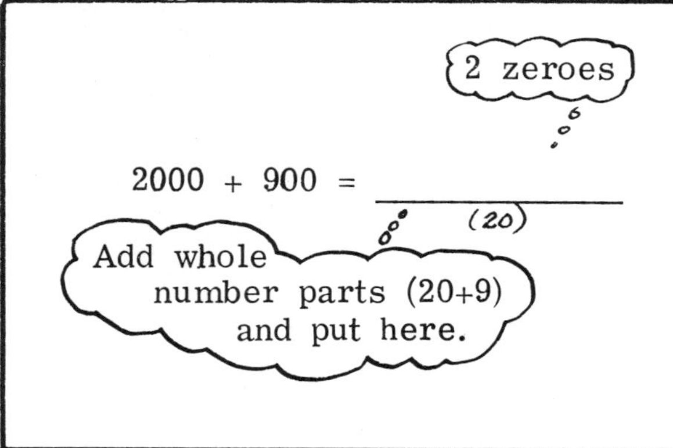

★ $300 + 8000 + 600 =$ _________ (21)

★ $50 + 100 + 90 =$ _________ (22)

★ $700 + 7000 + 600 =$ _________ (23)

Headquarters

General G. I. Dunno just arrived at Army Camp and cannot find Headquarters. To help him, estimate the answer to each problem below. Then follow the answers in order through the maze starting with answer #1.

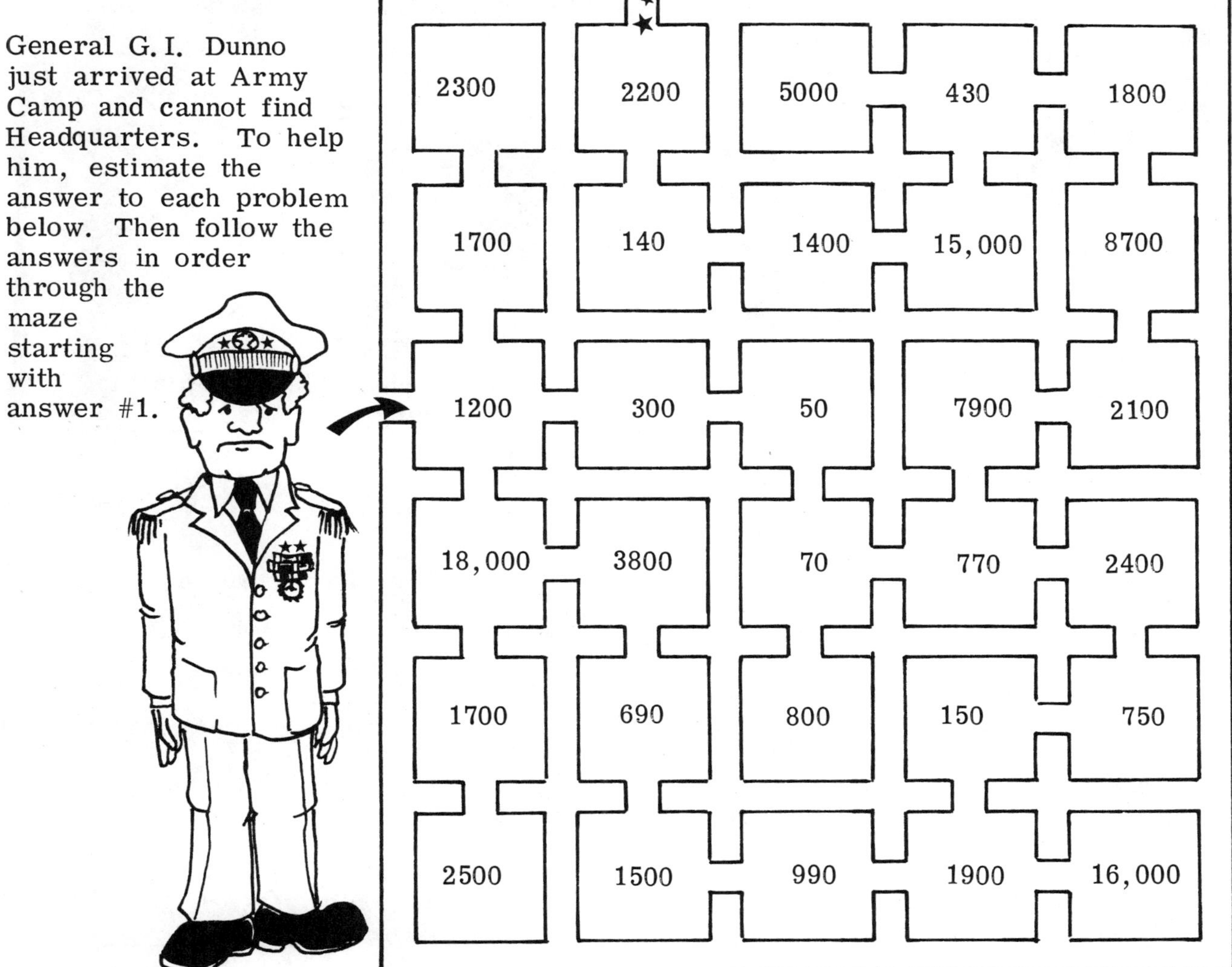

1. 300 + 400 + 500 = __1200__

2. 4000 + 6000 + 8000 = ______

3. 2000 + 1500 + 300 = ______

4. 70 + 600 + 20 = ______

5. 400 + 500 + 600 = ______

6. 100 + 90 + 800 = ______

7. 500 + 700 + 700 = ______

8. 30 + 50 + 70 = ______

9. 300 + 50 + 400 = ______

10. 700 + 800 + 900 = ______

11. 400 + 300 + 70 = ______

12. 2000 + 900 + 5000 = ______

13. 600 + 700 + 800 = ______

14. 3000 + 5000 + 700 = ______

15. 800 + 900 + 100 = ______

16. 90 + 300 + 40 = ______

17. 3000 + 5000 + 7000 = ______

18. 300 + 400 + 700 = ______

19. 30 + 40 + 70 = ______

20. 500 + 900 + 800 = ______

Making Some Headway   4

58,000 - 6000 = ?

What are the whole number parts of the numbers?

______ and ______
(24)   (25)

Does each number contain the same number of 0's? ______
(26)

How many? ______
(27)

1 Subtract the whole number parts in your head.

$$58000$$
$$- \ 6000$$

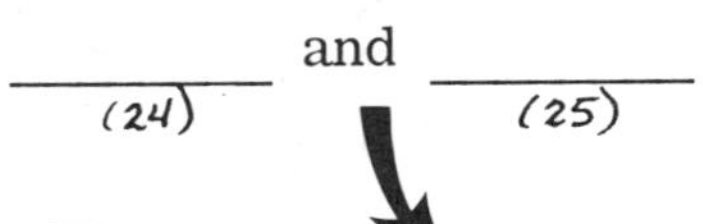
(28) ______ ______ (29)

2 Then put the same number of 0's on the answer.

2200 - 1200 = ?

1 Subtract the whole number parts in your head.

$$2200$$
$$- \ 1200$$

______ ______
(30)   (31)

2 Then put the same number of 0's on the answer.

Subtract in your head!

50 - 30 = ______
(32)

500 - 300 = ______
(33)

5000 - 3000 = ______
(34)

1800 - 700 = ______
(35)

47,000 - 37,000 = ______
(36)

2300 - 1500 = ______
(37)

80 - 50 = ______
(38)

1000 - 900 = ______
(39)

800 - 90 = ______
(40)

5200 - 5000 = ______
(41)

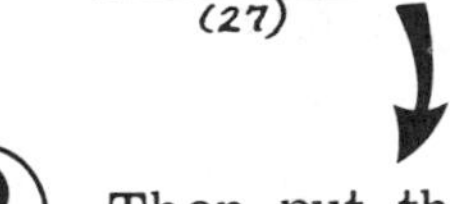

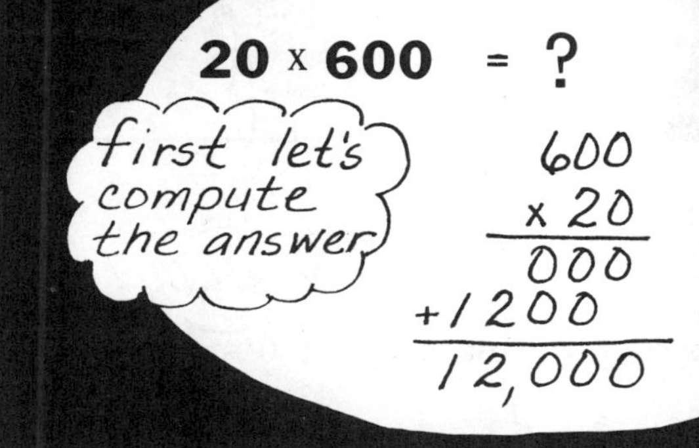

$20 \times 600 = ?$

Not using any zeroes,
what are the whole number
parts of the numbers?
______ and ______
(42)   (43)

What is the <u>total</u> number of
0's in <u>both</u> numbers? ______
(44)

(1) Multiply these to get
the whole number part
of the answer. ______
(45)

(2) Put this total number
of 0's on the answer.

★ Is this the same as the computed answer above? ______
(46)

Now try this: $1100 \times 6000 =$ ______
(47)

Multiply in your head.

$50 \times 70 =$ ______
(48)

$200 \times 80 =$ ______
(50)

$400 \times 600 =$ ______
(49)

$300 \times 12,000 =$ ______
(51)

$2800 \div 70 = \;?$

Not using any zeroes, what are the whole number parts of the numbers?

__________ and __________
(52)   (53)

(1) <u>Divide</u> the whole number parts in your head.

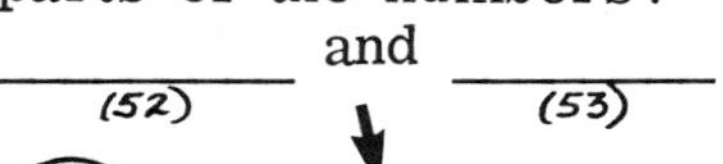

__________ (55)

<u>Subtract</u> the number of 0's.

$2 - 1 =$ __________ zero
(54)

(2) Put this number of 0's on the answer.

Divide in your head.

$640{,}000 \div 200 = \underline{3200}$

$15{,}000 \div 50 = \underline{00}$
(56)

$2400 \div 40 = $ __________
(57)

$480{,}000 \div 120 = $ __________
(58)

$5500 \div 500 = $ __________
(59)

$36{,}000 \div 60 = $ __________
(60)

A MINDLESS PUZZLE

Who _can't_ do this puzzle?

To answer this question, work each problem. Then find each answer in the Code Box and put the corresponding letter above the problem number in the blanks below. The first one is done for you.

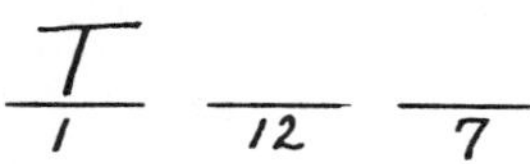

$$\frac{T}{1} \quad \overline{12} \quad \overline{7}$$

$$\overline{4} \quad \overline{13} \quad \overline{8} \quad \overline{16} \quad \overline{5} \quad \overline{11} \quad \overline{18} \quad \overline{18}$$

$$\overline{17} \quad \overline{3} \quad \overline{9} \quad \overline{6} \quad \overline{14} \quad \overline{10} \quad \overline{2} \quad \overline{15}$$

1. $8000 \div 40 = \underline{200}$

2. $80 \times 300 = \underline{\quad}$

3. $260,000 \div 130 = \underline{\quad}$

4. $5800 \div 20 = \underline{\quad}$

5. $40 \times 40 = \underline{\quad}$

6. $4600 \div 20 = \underline{\quad}$

7. $800 \div 40 = \underline{\quad}$

8. $70 \times 300 = \underline{\quad}$

9. $120 \times 30 = \underline{\quad}$

10. $1800 \div 60 = \underline{\quad}$

11. $360 \div 30 = \underline{\quad}$

12. $200 \times 50 = \underline{\quad}$

13. $800 \times 300 = \underline{\quad}$

14. $300 \times 110 = \underline{\quad}$

15. $9000 \div 30 = \underline{\quad}$

16. $16,000,000 \div 400 = \underline{\quad}$

17. $800 \times 400 = \underline{\quad}$

18. $150 \div 50 = \underline{\quad}$

Code Box

240,000	E
10,000	H
24,000	A
2000	O
33,000	E
230	S
1600	L
20	E
21,000	A
320,000	H
3	S
30	M
300	N
12	E
290	H
200	T
3600	R
40,000	D

1. 11
2. 6
3. 8
4. yes
5. 2
6. 25 } 2500
7. 00
8. 5
9. 13
10. 2
11. 1
12. 3
13. 21 } 21,000
14. 000
15. 35,000
16. 150
17. 2200
18. 1500
19. 21,000
20. 2900
21. 8900
22. 240
23. 8300
24. 58
25. 6

26. yes
27. 3
28. 52 } 52,000
29. 000
30. 10 } 1000
31. 00
32. 20
33. 200
34. 2000
35. 1100
36. 10,000
37. 800
38. 30
39. 100
40. 710
41. 200
42. 2
43. 6
44. 3
45. 12,000
46. yes
47. 6,600,000
48. 3500
49. 240,000
50. 16,000

51. 3,600,000

52. 28

53. 7

54. 1

55. 40

56. 300

57. 60

58. 4000

59. 11

60. 600

PUZZLE ANSWERS

page 4   (Maze Puzzle)

   it is self correcting

page 8   (Message Puzzle)

   THE HEADLESS HORSEMAN

gmp general math project

minneapolis public schools

UNIT: MAKING AN ADJUSTMENT

DESCRIPTION: In this unit, students learn that adjusted numbers are often
more convenient to use than exact numbers. They then
learn two methods by which to adjust numbers: truncation
and left-end rounding. A comparison of the two methods
is also made.

SKILLS TO BE REINFORCED: truncating
                           left-end rounding

PREREQUISITE SKILLS: none

LENGTH: 6 pages; 2 - 3 days

TEACHER NOTES: This is the third unit in the Estimation series. It is
recommended that they be used in sequence.

NO TEST

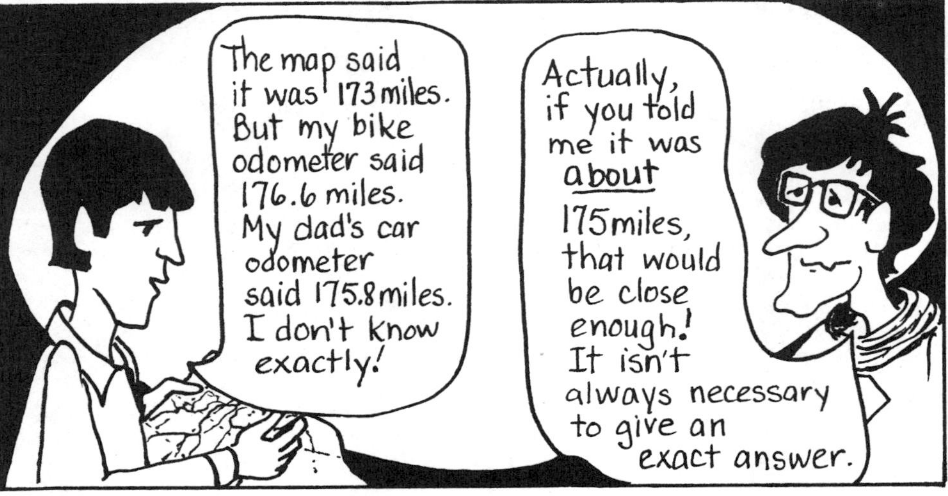

Estimating involves <u>adjusting</u> exact numbers to get numbers that are easier to work <u>with</u>, but are still close to the exact numbers.

There are several ways to <u>adjust</u> numbers. One way is to TRUNCATE them.

To TRUNCATE a number, leave the first digit as it is and replace the other digits with 0's.

These are the exact numbers

8172 is truncated to 8000

32 is truncated to 30

27,653 is truncated to 20,000

These are the truncated numbers

TRUNCATING GIVES A RESULT THAT...

- has one digit followed by 0's
- is quite easy to work with

✷ Adjust these numbers by truncating.

1278	__________ (1)	169,873	__________ (4)	13,467	__________ (7)
439	__________ (2)	959	__________ (5)	1892	__________ (8)
27	__________ (3)	602	__________ (6)	43	__________ (9)

✷ Write <u>three</u> numbers that truncate to the given number.

30: _____38_____ , __________ , __________ (10)

400: __________ , __________ , __________ (11)

80: __________ , __________ , __________ (12)

200: __________ , __________ , __________ (13)

6000: __________ , __________ , __________ (14)

✷ Now do these problems;
adjust numbers using <u>truncation</u>!

The exact answer is $32 x 52.

Truncating, this is about $30 x 50 = $ __________ . (15)

The exact answer is $8289 - __________ . (16)

Truncating, this is about __________ - __________ = __________ . (17) (18) (19)

The exact answer is __________ + __________ + __________ . (20) (21) (22)

This is about $ 20 + __________ + __________ = __________ . (23) (24) (25) (26)

Making an Adjustment   2

8962 truncates to 8000, but it is closer to 9000.

574 truncates to 500, but it is closer to 600.

17 truncates to 10, but it is closer to 20.

★ 28 truncates to _______, but it is closer to 30.
        (27)

692 truncates to _______, but it is closer to _______.
        (28)   (29)

1602 truncates to _______, but it is closer to _______.
        (30)   (31)

Sometimes you want an adjusted number that is really easy to work with but is closer to the exact number than the truncated number might be. Then use LEFT-END ROUNDING to get the adjusted number.

LEFT-END ROUNDING (LER) GIVES A RESULT THAT ...

● has one digit followed by 0's

● is easy to work with

● is closer to the exact number than the truncated result might be

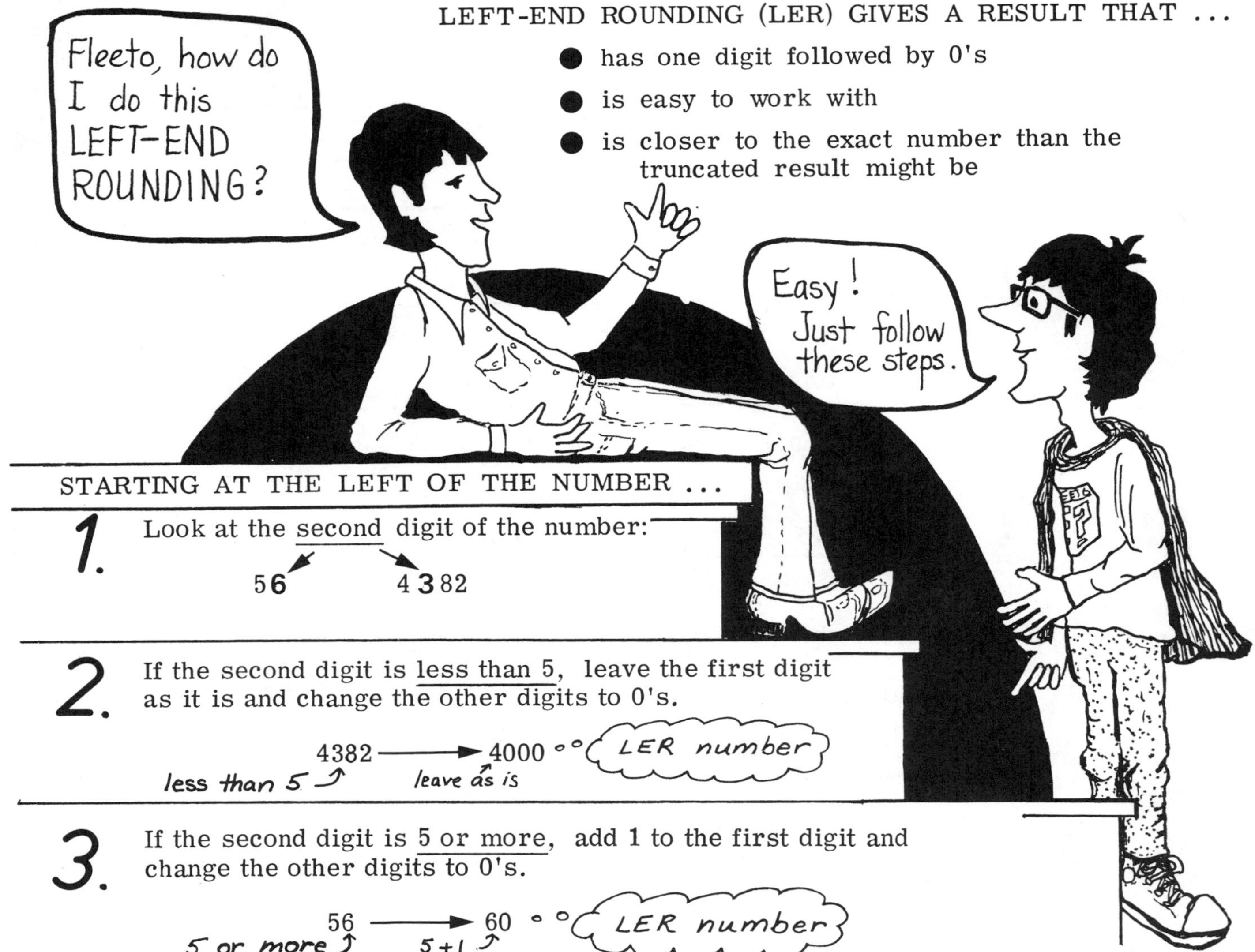

     Making an Adjustment 3

✴ Adjust these numbers using Left-End Rounding:

127 __________ (32)   567 __________ (35)   9820 __________ (38)

48 __________ (33)   33 __________ (36)   17,520 __________ (39)

2473 __________ (34)   471 __________ (37)   64 __________ (40)

A `CHEEP` SHOT

What do you call a bird that over-estimated what it should eat?

___ ___ ___ ___ ___ ___   ___ ___ ___ ___ ___
 1   11 9 8 2 5   10 6 7 4 3

Use Left-End Rounding to adjust each number below. Then find each answer in the Code Box and put the corresponding letter above the problem number.

(1) 89 __________   (7) 62 __________

(2) 621 __________   (8) 538 __________

(3) 409 __________   (9) 17 __________

(4) 6581 __________   (10) 651 __________

(5) 27 __________   (11) 75 __________

(6) 6090 __________

O	D	B	I	R	A	N	U	N	R	O	I
20	30	60	70	80	90	400	500	600	700	6000	7000

                    Making an Adjustment   4

HEAVY INTO ESTIMATING

Adjust each number below. Then shade each area that contains an answer. The first two are done for you.

Truncate	3485	*3000*
LER	4769	*5000*
Truncate	56	
Truncate	6481	
LER	74	
Truncate	409	
LER	471	
Truncate	42	
LER	953	

Truncate	962	
LER	137	
Truncate	626	
LER	328	
LER	3788	
Truncate	36	
LER	19	
Truncate	19	
LER	1699	

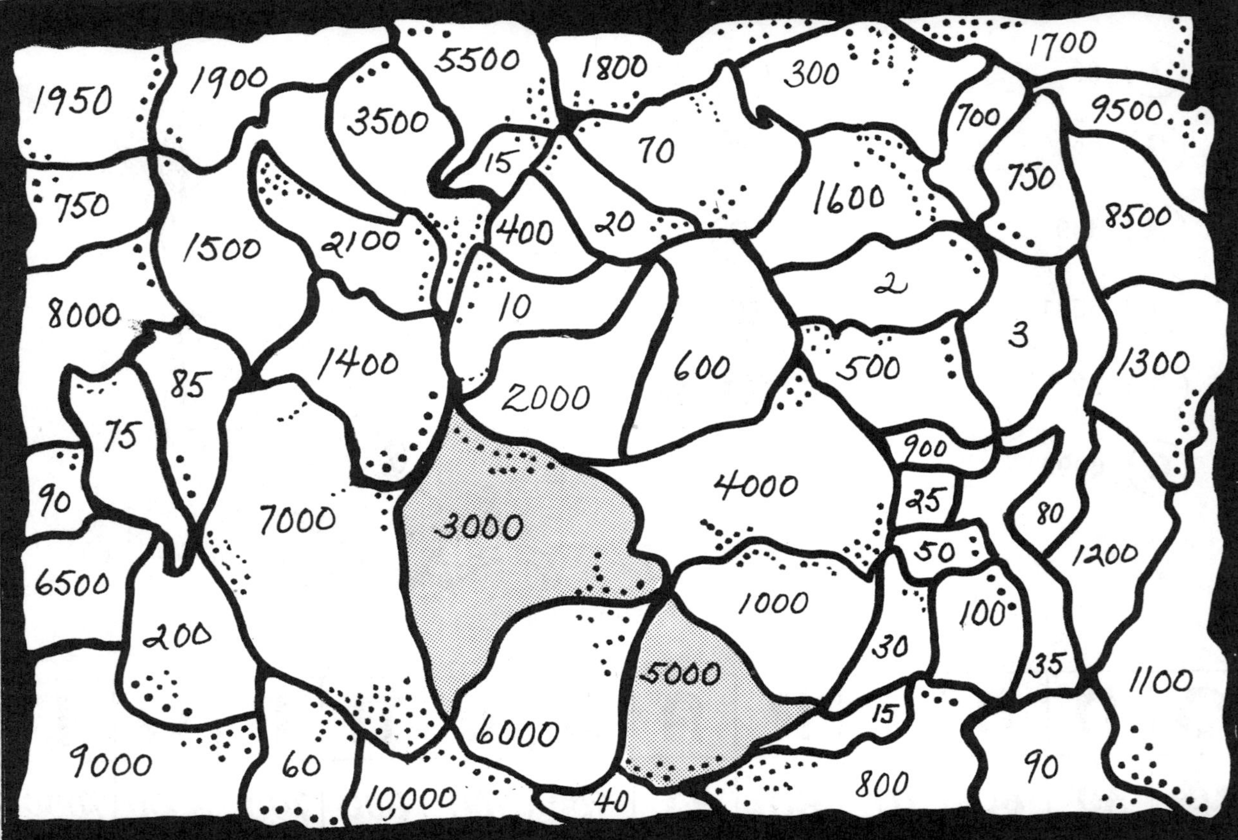

1. Write a whole number that truncates to:

50 _________ (41)   400 _________ (42)   8000 _________ (43)   20,000 _________ (44)

2. Write a whole number that LER's to:

20 _________ (45)   80,000 _________ (46)   1000 _________ (47)   200 _________ (48)   30 _________ (49)

3. Write <u>all</u> the numbers that....

truncate to 30 _________ (50)   LER to 30 _________ (51)

truncate to 400 _________ (52)   LER to 400 _________ (53)

4. Describe a number for which the truncated and LER results will be the same:

_________ (54)

Describe a number for which the truncated and LER results will differ:

_________ (55)

5. A new car costs $9792.

The truncated cost is $ _________ (56); the LER cost is $ _________ (57).

Which adjusted number is closer to the actual cost? _________ (58).

Which adjusted number is more useful? _________ (59).

Why? _________ (60)

6. The diameter of the earth at the equator is 7927 miles.

Truncating, this is _________ (61); LER gives _________ (62).

Which adjusted number is closer to the actual diameter? _________ (63)

Which adjusted number would be most useful? _________ (64)

Why? _________ (65)

7. Which generally gives the most useful result: truncating or LER ? _________ (66)

Why? _________ (67)

1. 1000
2. 400
3. 20
4. 100,000
5. 900
6. 600
7. 10,000
8. 1000
9. 40
10. any 2 different numbers between 30 and 39, except 38
11. any 3 different numbers between 400 and 499
12. any 3 different numbers between 80 and 89
13. any 3 different numbers between 200 and 299
14. any 3 different numbers between 6000 and 6999
15. 1500
16. 6347
17. $8000
18. $6000
19. $2000
20. $26.85
21. $12.69
22. $33.18
23. $20
24. $10
25. $30

26. $60
27. 20
28. 600
29. 700
30. 1000
31. 2000
32. 100
33. 50
34. 2000
35. 600
36. 30
37. 500
38. 10,000
39. 20,000
40. 60
41. any number between 50 & 59
42. any number between 400 & 499
43. any number between 8000 & 8999
44. any number between 20,000 & 29,999
45. any number between 15 & 24
46. any number between 75,000 & 84,999
47. any number between 950 & 1499
48. any number between 150 & 249
49. any number between 25 & 34
50. 30 . . . 39

51. 25 . . . 34

52. 400 . . . 499

53. 350 . . . 449

54. numbers where the second digit is less than 5.

55. numbers where the second digit is ≥ 5

56. 9000

57. 10,000

58. LER or $10,000

59. LER or $10,000

60. it is closer to the actual cost and easier to use

61. 7000

62. 8000

63. 8000 or LER

64. 8000

65. it is closer to the actual diameter and easier to use

66. LER

67. it is closer to the actual answer

PUZZLE ANSWERS

page 4   (Message Puzzle)
A ROUND ROBIN

page 5   (Shading Puzzle)
it is an elephant

   Making an Adjustment   A2

 **_general math project_**

minneapolis public schools

UNIT: 'JUST AROUND THE CORNER

DESCRIPTION: This unit introduces rounding to the nearest 10, 100
and 1000. Rounded numbers are compared to Truncated
and Left-End Rounded numbers.

SKILLS TO BE REINFORCED: problem solving skills

rounding

PREREQUISITE SKILLS: truncating

left-end rounding

LENGTH: 8 pages; 2 - 3 days

TEACHER NOTES: The students will see that the different methods of
adjusting numbers give different answers, each with
its own use. Rounded results are usually close to
the exact numbers.

NO TEST

"JUST AROUND THE CORNER

Freddy's Fast Foods

Hamburger.....	$ .59		French Fries	
Basket	1.49		Small	$ .69
Cheeseburger ..	.79		Large	.89
Basket	1.69		Chocolate Shake ...	.69
Cole Slaw	.47		Milk	.49
Apple	.35			

The actual cost of the lunch is $1.49 + $.69 + $.35

Truncating, this is about $1.00 + $.60 + $.30 = __________ . (1)

LER gives about $1.00 + __________ + __________ = __________ . (2) (3) (4)

Should $2.40 be enough for lunch? __________ (5)

Let's estimate the cost of the lunch again, but adjust the prices by ROUNDING TO THE NEAREST 10¢.

Hamburger basket .... $1.50 ←

Chocolate shake .... .70 ←

Apple .... .40 ←

ESTIMATED TOTAL ☐
(6)

Using the estimated total, would you decide that $2.40 is enough for lunch? ________
(7)

Why does rounding to the nearest 10 give a better estimate than truncating or LER?

____
(8)

ROUNDING TO THE NEAREST 10

To the nearest 10 ...

41 rounds to **40**

783 rounds to **780**

6429 rounds to **6430**

95 rounds to **100**

TO ROUND A NUMBER TO THE NEAREST 10...

Look at the <u>last</u> digit of the number.

If it is <u>less than 5</u>, change it to 0 and leave the other digits.

If it is <u>5 or more</u> change it to 0, add 1 to the digit on its left, and leave the other digits.

192 → 190

157 → 160

✴ Round to the nearest 10:

1969   **19_0** (9)   2341   **23_0** (10)   8919 ______ (11)   83 ______ (12)

18,723 ______ (13)   59 ______ (14)   377 ______ (15)   124 ______ (16)

✴ Round 498 to the nearest 10. ______ (17)

Sometimes rounding to the nearest 10 isn't as useful as ROUNDING TO THE NEAREST 100.

$5427 could be adjusted...

by truncating or LER to $5000.   This is an easy number with which to compute, but it is not very close to $5427.

by rounding to the nearest 10 to $5430.   This is close to $5427, but would not be easy to work with in your head.

by ROUNDING TO THE NEAREST 100 to $5400.   This is close to $5427 and is fairly easy to work with in your head.

ROUNDING TO THE NEAREST 100

To the nearest 100 ...

232 rounds to **200**

4656 rounds to **4700**

95 rounds to **100**

TO ROUND A NUMBER TO THE NEAREST 100 ...

Look at the number formed by the last two digits.

If it is less than 50, change these two digits to 0's and leave the other digits.

637 ➡ **600**
less than 50 · leave as is · Change to 0's

If it is 50 or more, change these two digits to 0's, add 1 to the next digit on the left and leave the other digits.

1267 ➡ **1300**
50 or more · leave · add 1 · Change to 0's

✱ Round to the nearest 100:

448	__OO__ (18)	6515	__OO__ (19)	1974	______ (20)
14,892	______ (21)	81	______ (22)	297	______ (23)

HEAVY SETTER
My dog is fat, so I named him....

__	_O_	__	__	__		__	__	__	__	__
4	3	10	7	1		8	2	6	5	9

Round each number on the left to the nearest 100. Then draw a line from the number to its corresponding rounded number on the right. Each line will pass through a letter and a number. Put the letter above the number in the message. The first one is done for you.

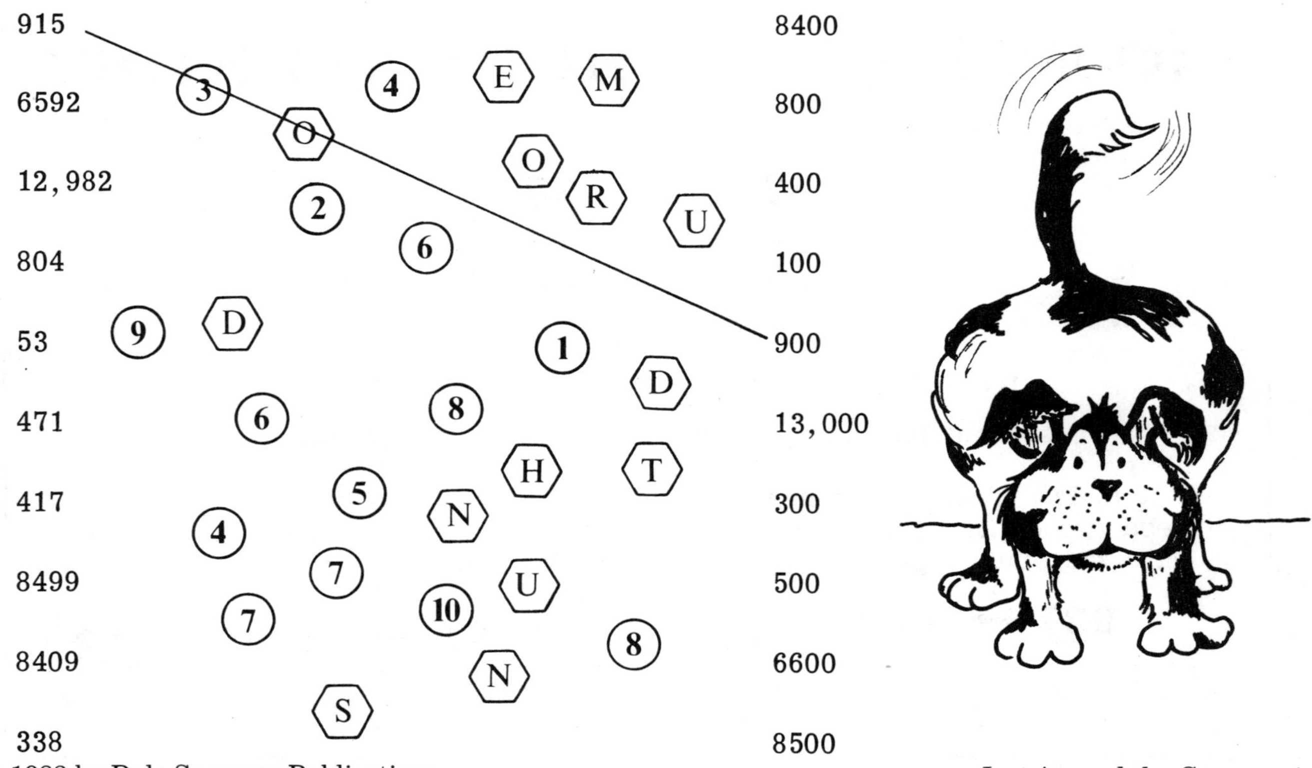

Light travels at 186,000 miles per second. This is not an exact number. It is an adjusted number that is close enough to use for most of our purposes. It is an adjusted number that is ROUNDED TO THE NEAREST 1000.

ROUNDING TO THE NEAREST 1000

To the nearest 1000 ...

3267 rounds to **3000**

17,555 rounds to **18,000**

978 rounds to **1000**

TO ROUND A NUMBER TO THE NEAREST 1000 ...

Look at the number formed by the last three digits.

If it is less than 500, change these three digits to 0's and leave the other digits.

If it is 500 or more, change these three digits to 0's, add 1 to the next digit on the left and leave the other digits.

2225 → 2000

3745 → 4000

THE ROUND TABLE

To find out what the doodle is, round each number to the nearest 1000. Then find the rounded number in the Code Box and put the corresponding letter above the problem number.

1. 4128 __________

2. 15,090 __________

3. 28,912 __________

4. 6112 __________

5. 33,333 __________

6. 90,909 __________

7. 8713 __________

8. 7103 __________

9. 83,050 __________

10. 2484 __________

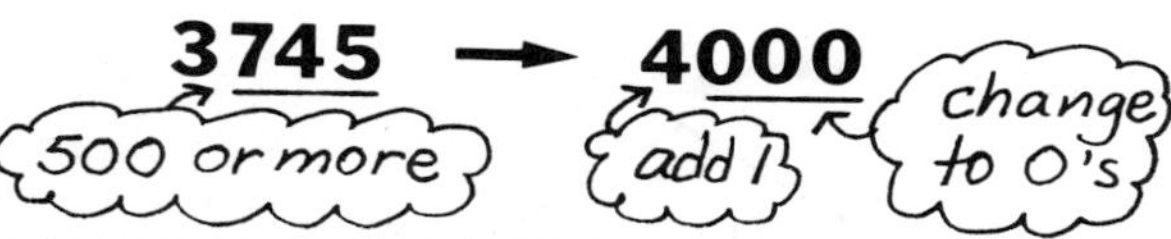

___ ___ ___ ___ ___ ___ ___ ___ ___ ___
10 6 2 5 9 3 1 8 4 7

Code Box

29,000	(N)	7000	(G)	4000	(I)	2000	(M)
9000	(T)	15,000	(D)	19,000	(P)	33,000	(—)
3000	(W)	91,000	(I)	83,000	(K)	6000	(H)

Estimation is like this. Depending on how you adjust the numbers, you can get different estimated answers for the same problem. You have to decide what approach is best for each individual situation.

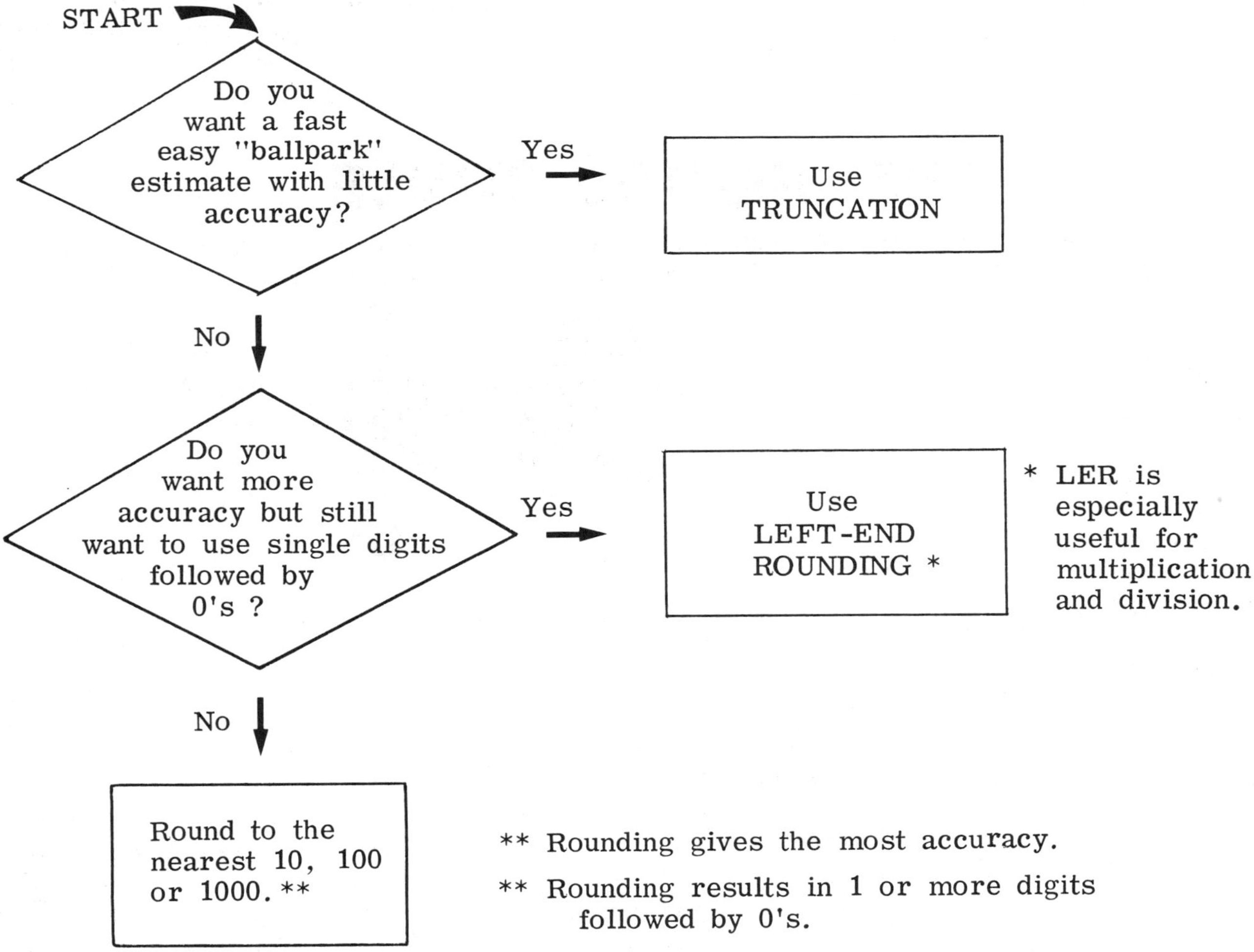

START

Do you want a fast easy "ballpark" estimate with little accuracy?

Yes → Use TRUNCATION

No ↓

Do you want more accuracy but still want to use single digits followed by 0's ?

Yes → Use LEFT-END ROUNDING *

* LER is especially useful for multiplication and division.

No ↓

Round to the nearest 10, 100 or 1000. **

** Rounding gives the most accuracy.

** Rounding results in 1 or more digits followed by 0's.

Here are some examples of rounded numbers:

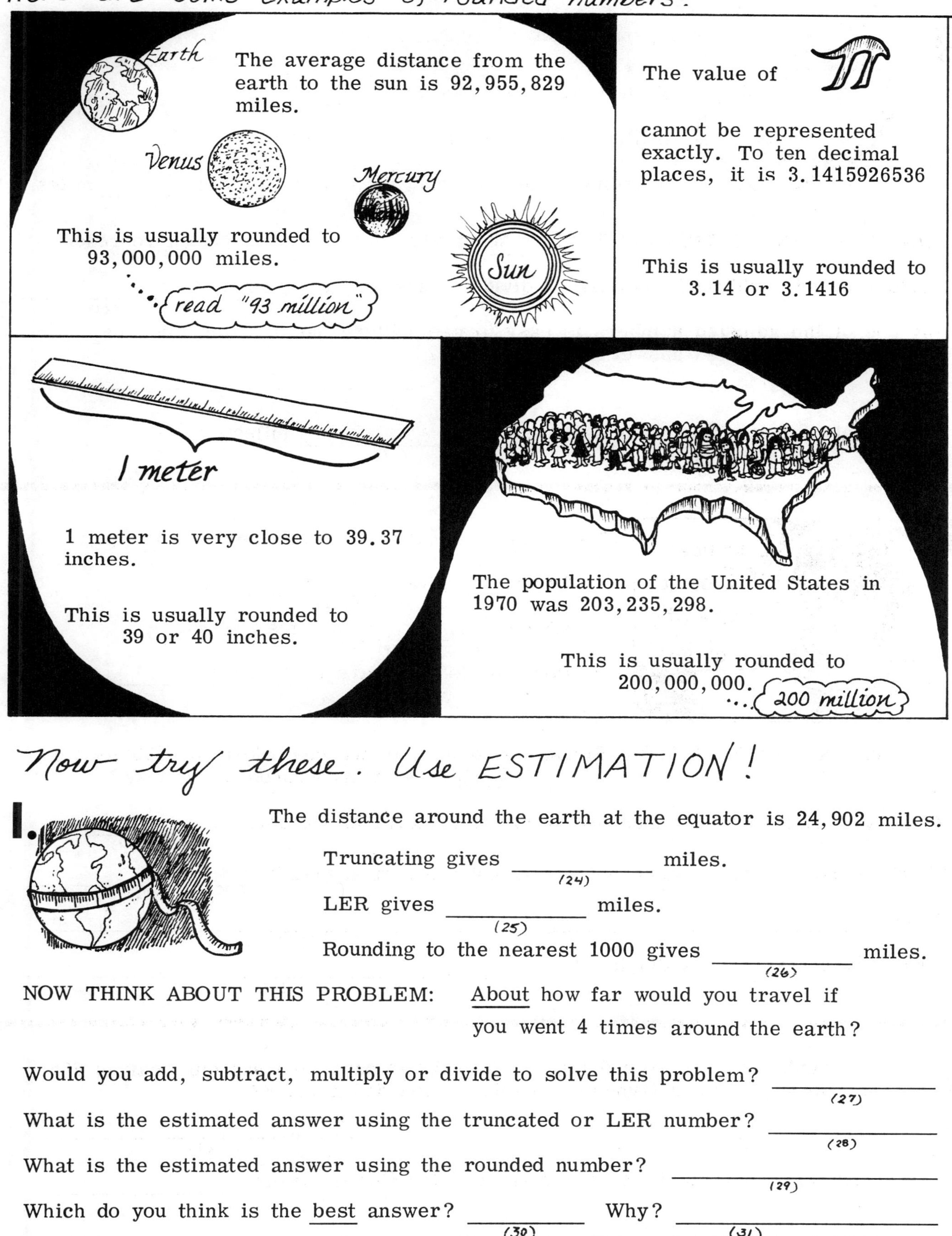

Now try these. Use ESTIMATION!

I. The distance around the earth at the equator is 24,902 miles.

Truncating gives _____________ miles.
(24)

LER gives _____________ miles.
(25)

Rounding to the nearest 1000 gives _____________ miles.
(26)

NOW THINK ABOUT THIS PROBLEM: <u>About</u> how far would you travel if you went 4 times around the earth?

Would you add, subtract, multiply or divide to solve this problem? _____________
(27)

What is the estimated answer using the truncated or LER number? _____________
(28)

What is the estimated answer using the rounded number? _____________
(29)

Which do you think is the <u>best</u> answer? _____________   Why? _____________
(30)   (31)

2. On a certain map of Texas, 1 inch = 58 miles.

Truncating, 1 inch is about _______ miles.
(32)

LER, 1 inch is about _______ miles.
(33)

Rounding to the nearest 10, 1 inch is about _______ miles.
(34)

NOW THINK ABOUT THIS PROBLEM: <u>About</u> what distance does $\frac{1}{2}$ inch represent?

Would you add, subtract, multiply or divide to answer this question? _______
(35)

Any one of the adjusted numbers is <u>easy</u> to use. Which one(s) would give the <u>closest</u> estimated answer to the question?

____
(36)

About what distance does $\frac{1}{2}$ inch represent? _______ miles.
(37)

3.

The speed of sound is 1088 feet per second (fps).

Truncating gives _______ fps.
(38)

LER gives _______ fps.
(39)

Rounding to the nearest 100 gives _______ fps.
(40)

NOW THINK ABOUT THIS PROBLEM: A sound has traveled close to 3300 feet. <u>About</u> how many seconds did it take?

Would you add, subtract, multiply or divide to solve this problem? _______
(41)

Which adjusted number would you use to get the best answer? _______
(42)

Why? _______
(43)

About how many seconds did it take? _______
(44)

4. The earth revolves around the sun in 365.244 days. We say our year is 365 days long. Is this a truncated, LER, or rounded number?

____
(45)

What would happen if we used LER to determine our year? _______

____

                Just Around the Corner   8

1. $1.90
2. $ .70
3. $ .40
4. $ 2.10
5. yes
6. $ 2.60
7. no
8. closer to actual cost
9. 1970
10. 2340
11. 8920
12. 80
13. 18,720
14. 60
15. 380
16. 120
17. 500
18. 400
19. 6500
20. 2000
21. 14,900
22. 100
23. 300
24. 20,000
25. 20,000

26. 25,000
27. multiply
28. 80,000
29. 100,000
30. 100,000
31. closer to actual answer
32. 50
33. 60
34. 60
35. divide or multiply
36. LER and rounding
37. 30
38. 1000
39. 1000
40. 1100
41. divide
42. 1100
43. closer to actual answer
44. 3
45. rounded number
46. we would have 400 days per year (too many days)

PUZZLE ANSWERS

page 4 (Message Puzzle)

ROUND HOUND

page 5 (Message Puzzle)

MID-KNIGHT

     Just Around the Corner   A

 **general math project**

minneapolis public schools

UNIT: BE REASONABLE

DESCRIPTION: In this unit the student will learn to use estimation
to check the reasonableness of typical classroom
problems. This includes addition, subtraction,
multiplication, and division of whole numbers,
multiplication and division of decimals and calculator
work.

SKILLS TO BE REINFORCED: alertness to reasonableness of answers
adjusting numbers
mental arithmetic
problem solving

PREREQUISITE SKILLS: addition, subtraction, multiplication and
division of whole numbers and decimals

LENGTH: 9 pages;   2 - 3 days

TEACHER NOTES: You may want to review the unit MAKING SOME
HEADWAY before doing this unit, and then spend
some time discussing the ideas presented in this
unit. Checking answers for reasonableness
should be emphasized in many class assignments.

NO TEST

BE REASONABLE

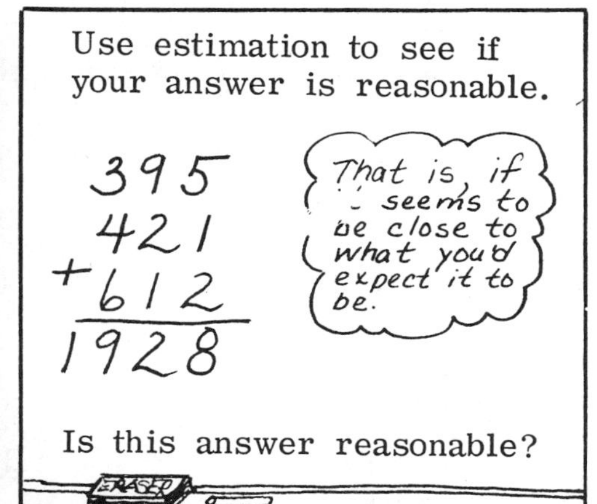

To answer this question . . .

ESTIMATE **395 + 421 + 612**

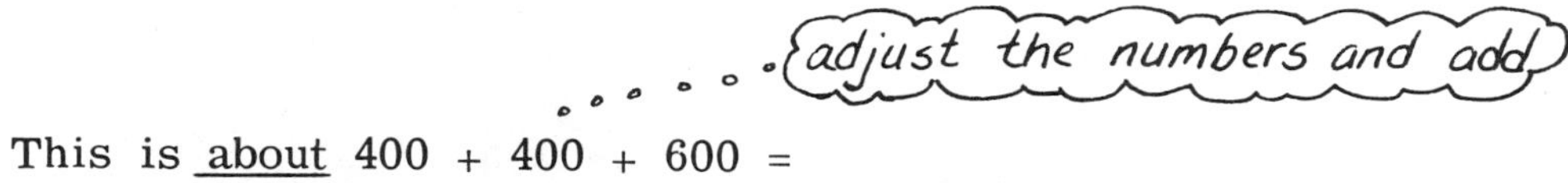

* This is <u>about</u> 400 + 400 + 600 = ________
  (1)

* Compare your estimate and the answer on the board above.

* Is 1928 a reasonable answer to 395 + 421 + 612 or should you redo the problem? ________________ ________
  (2)

* If you need to redo 395 + 421 + 612, what should the answer be? ________
  (3)

                    Be Reasonable 1

Your teacher puts these examples on the board.   Use ESTIMATION to tell if the answers are reasonable.

ESTIMATE **97 x 11**

✳ This is about 100 x _______ = _______
         (4)   (5)

✳ Compare your estimate and the answer on the board above.

✳ Is 167 a reasonable answer to 97 x 11 or should you redo the problem? _______
         (6)

✳ If you need to redo 97 x 11, what is the answer? _______
                                                        (7)

ESTIMATE **148 - 34**

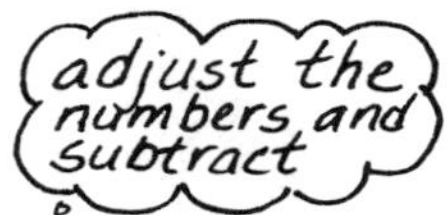

✲ This is about _______ - _______ = _______
                  (8)   (9)   (10)

✲ Compare your estimate and the answer on the board above.

✲ Is 114 a reasonable answer to 148 - 34 or should you redo the problem? _______
                                                                            (11)

✲ If you need to redo 148 - 34, what would the answer be? _______
                                                              (12)

ESTIMATE **4297 + 3263 + 1801**

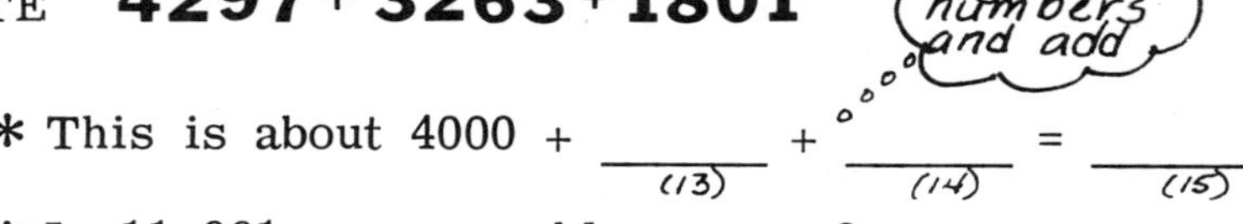

✲ This is about 4000 + _______ + _______ = _______
                         (13)   (14)   (15)

✲ Is 11,661 a reasonable answer? _______
                                    (16)

✲ If not, what is 4297 + 3263 + 1801? _______
                                          (17)

                              Be Reasonable 2

Use estimation to help you do these problems:

1 Tell whether each answer is reasonable.   If not, redo the problem.

$$62$$
$$89$$
$$+\ 426$$
$$\overline{977}$$

Answer reasonable? _______ (18)

If not, redo: _______ (19)

$$64$$
$$\times\ 47$$
$$\overline{3008}$$

Answer reasonable? _______ (20)

If not, redo: _______ (21)

---

324 - 275 = 129

Answer reasonable? _______ (22)

If not, redo: _______ (23)

195 x 33 = 8435

Answer reasonable? _______ (24)

If not, redo: _______ (25)

2 Which one has the <u>wrong</u> answer? Circle it.

(32 x 59) + 19 = 807

(87 x 9) + 42 = 825   (26)

3 Which answer is the <u>largest</u>? Circle it.

63 x 49

19 x 98

52 x 52

(27)

4 Which answer is <u>smallest?</u> Circle it.

414 - 188

531 - 178

685 - 567

(28)

5 Which number would give the answer to: ☐ + 87 = 118

a) 12

b) 31

c) 54

(29)

6 Which number would give the answer to: 19 x ☐ = 361

a) 19   b) 36   c) 7

(30)

          Be Reasonable 3

To get an estimated answer to a division problem...

1 Adjust the divisor to get a single digit followed by zeroes.

2 Cross out any zeroes in the divisor, leaving a single digit; count the number of zeroes you crossed out and cross out the same number of digits in the dividend.

3 Find the first digit in the answer. Fill it in. Then fill in the remaining places with zeroes. Put one zero above every digit in the dividend.

60 is the estimated answer.

So, to be reasonable, the answer to 67)4293 should be about 60.

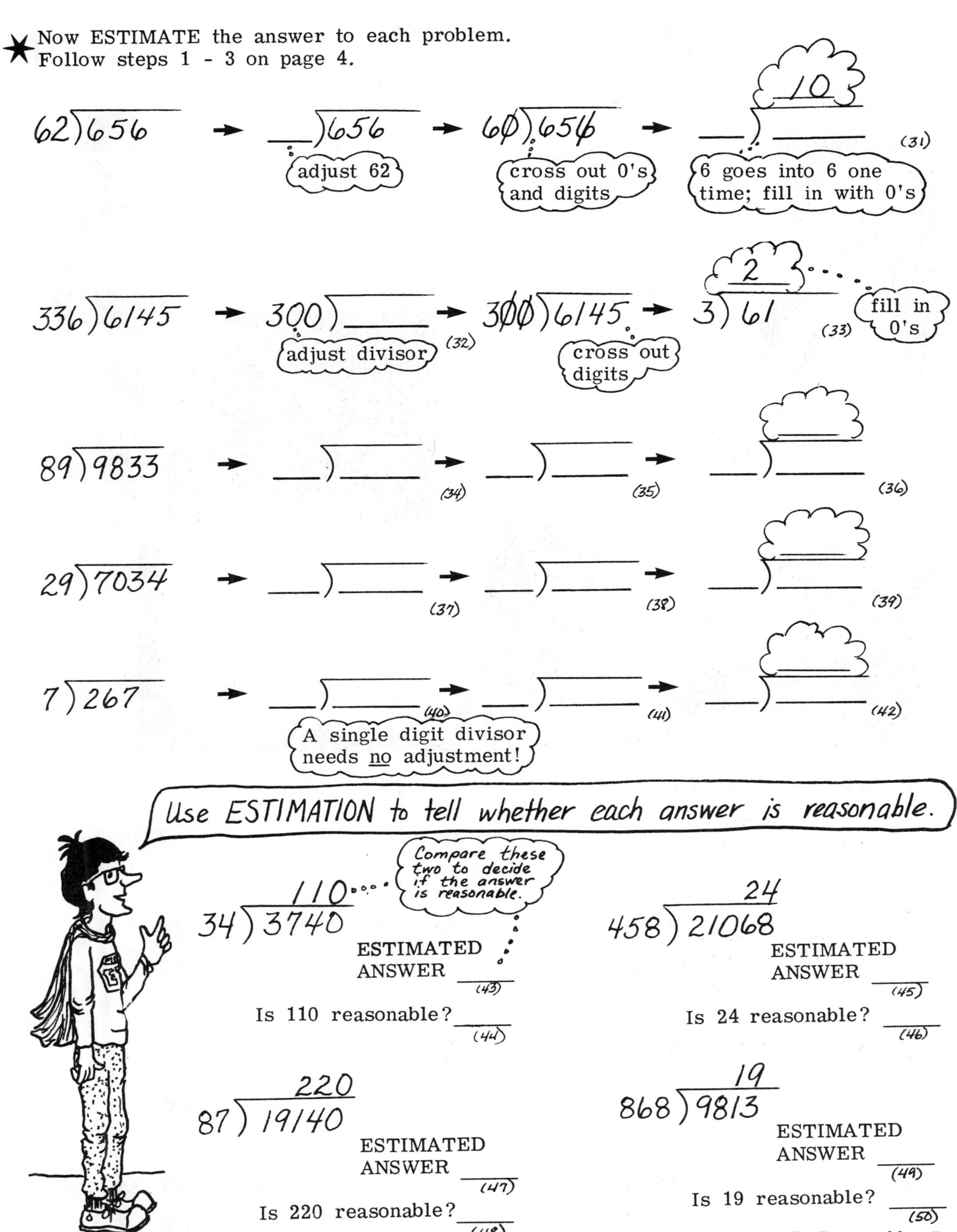

Be Reasonable 5

1. 4.8 x 5.1 = 2448

This is about 5 x 5.

(51)

2. 23.9 x 7.6 = 18164

This is about 20 x ___

(52)

3. 581.6 x 1.1 = 63976

(53)

4. 6.24 x 41.6 = 259584

(54)

5. 1.95 x 33 = 6435

(55)

6. 8.82 x 27 = 23814

(56)

7. 95 x 4.6 = 4370

(57)

8. 10.67 x 2.601 = 2775267

(58)

1. $58.5 \div 2.5 = 234$

   This is about $60 \div 3$ (59)

2. $41.472 \div 5.12 = 81$

   This is about $40 \div 5$ (60)

3. $218.94 \div 26.7 = 82$

   (61)

4. $200.56 \div 8.72 = 23$

   Don't let this one trick you. (62)

5. $12.16 \div 3.8 = 32$

   (63)

6. $2963.7 \div 8.9 = 333$

   (64)

7. $59.448 \div 1.2 = 4954$

   (65)

8. $93.67 \div 1.9 = 493$

   (66)

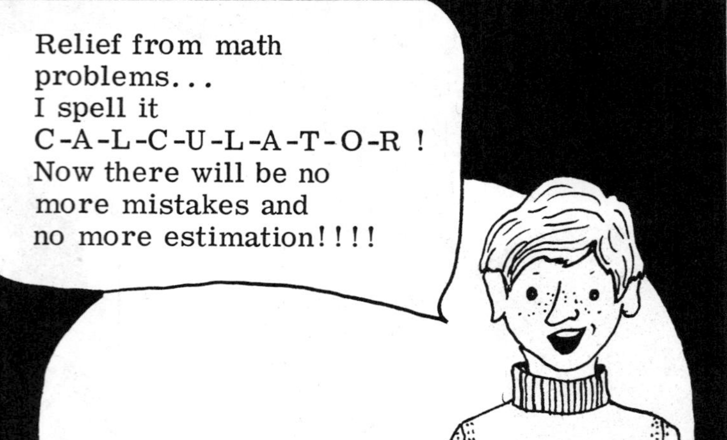

★ Try these! Adjust each number in the problem and then ESTIMATE to see if each answer is reasonable.

Don wants to buy a turntable for $199.95 and a tape deck for $127.50.   To find out how much money he needs altogether, he did this on his calculator:

$$\boxed{199.95} \ \boxed{+} \ \boxed{12.750} \ \boxed{=} \ \boxed{212.70}$$

Is this answer reasonable? _________ (67)

If not, what was his mistake? _________
_________________________________ (68)

A new roof on Tony's house will cost $4200, and a new foundation under the porch will cost $9779.   To find out what the total repair bill will be, he did this on his calculator:

$$\boxed{4200} \ \boxed{+} \ \boxed{9779} \ \boxed{=} \ \boxed{13979}$$

Is this answer reasonable? _________ (69)

If not, what was his mistake? _______
_________________________________ (70)

Roscoe is paying $120 per month, for 17 months, to pay for his car. To find out how much the car costs, he did this on his calculator:

| 1200 | x | 17 | = | 20400 |

Is this answer reasonable? _______ (71)

If not, what was his mistake? _______
_________________________________ (72)

A machine bolt weighs 42 grams. To find the weight of a box of 150 bolts, Anita did this on her calculator:

| 42 | + | 150 | = | 192 |

Is this answer reasonable? _______ (73)

If not, what was her mistake? _______
_________________________________ (74)

---

1 inch = 2.54 centimeters. To find out how many centimeters there are in a foot, Dan did this on his calculator:

| 12 | x | 2.54 | = | 30.48 |

Is this answer reasonable? _______ (75)

If not, what was his mistake? _______
_________________________________ (76)

---

1 meter = 39.37 inches. To find out how many inches there are in 15 meters, Bonnie did this on her calculator:

| 93.37 | x | 15 | = | 1405.95 |

Is this answer reasonable? _______ (77)

If not, what was her mistake? _______
_________________________________ (78)

Pat read a 792 page book in 12 days. To find out how many pages she averaged each day, she did this on her calculator:

| 792 | ÷ | 12 | = | 66 |

Is this answer reasonable? _______ (79)

If not, what was her mistake? _______
_________________________________ (80)

Janice is paying $12.00 per week on a bike that costs $179.95. To find out how many payments she must make, she did this on her calculator:

| 12.00 | ÷ | 179.95 | = | 0.0666851 |

Is this answer reasonable? _______ (81)

If not, what was her mistake? _______
_________________________________ (82)

1. 1400
2. redo the problem
3. 1428
4. 10
5. 1000
6. redo
7. 1067
8. 150
9. 30
10. 120
11. reasonable
12. OK as is
13. 3000
14. 2000
15. 9000
16. no
17. 9361
18. no
19. 577
20. yes
21. —
22. no
23. 49
24. no
25. 6435

26. (32×59)+19 =807
27. 63 × 49
28. 685 − 567
29. b) 31
30. a) 19
31. 6)65
32.)6145
33. 20
34. 90)9833
35. 90)9833
36. 9)983   (100)
37. 30)7034
38. 30)7034
39. 3)703   (200)
40. 7)267
41. 7)267
42. 7)267   (30)
43. 100
44. yes
45. 40
46. no
47. 200
48. yes
49. 10
50. no

51. 24.48
52. 181.64
53. 639.76
54. 259.584
55. 64.35
56. 238.14
57. 437.0
58. 27.75267
59. 23.4
60. 8.1
61. 8.2
62. 23.
63. 3.2
64. 333.
65. 49.54
66. 49.3
67. no
68. 12.750 should have been 127.50
69. yes
70. —
71. no
72. 1200 should have been 120
73. no
74. the "+" should have been "x"
75. yes

76. —
77. no
78. 93.37 should have been 39.37
79. yes
80. —
81. no
82. the problem should have been 179.95 ÷ 12.00

gmp general math project

minneapolis public schools

UNIT: DIVIDE AND CONQUER

DESCRIPTION: In this unit, students learn how estimation is used in the long division process.

SKILLS TO BE REINFORCED: rounding
estimating
division of whole numbers

PREREQUISITE SKILLS: subtraction, multiplication and division of whole numbers

LENGTH: 3 pages; 1 - 2 days

TEACHER NOTES: Most students have difficulty deciding what numbers to use in the quotient of long division. By using the newly mastered skills of rounding and estimating, they can overcome this old problem.

You might want to go over the first two pages of this unit with students and encourage a discussion about using estimation in division.

NO TEST

DIVIDE AND CONQUER

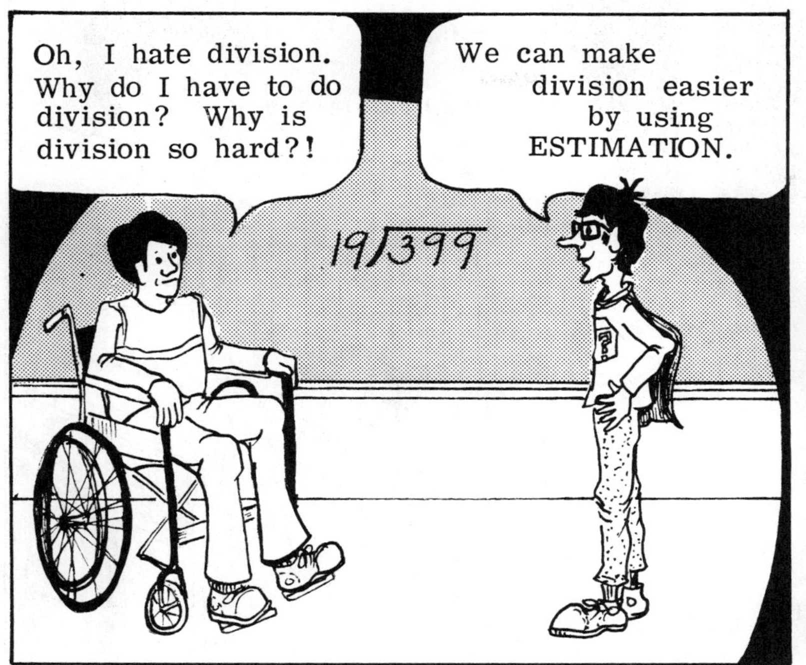

Look at the problem from the board: 19⟌399 *How do you do this?*

Let's try . . . **29)464̄**

29 doesn't go into 4, but it does go into 46.
29 <u>is about</u> 30; 30 goes into 46 once - twice
would be too much.

$$29\overline{)464} \quad \frac{1}{29}$$
$$\overline{174}$$

Now look at 29 into 174.   This <u>is about</u> 30
into 170. It could be 5 or 6, but it looks
closer to 6.
Try 6 first.

$$29\overline{)464} \quad \frac{16}{29}$$
$$\overline{174}$$
$$\underline{174}$$
$$0$$

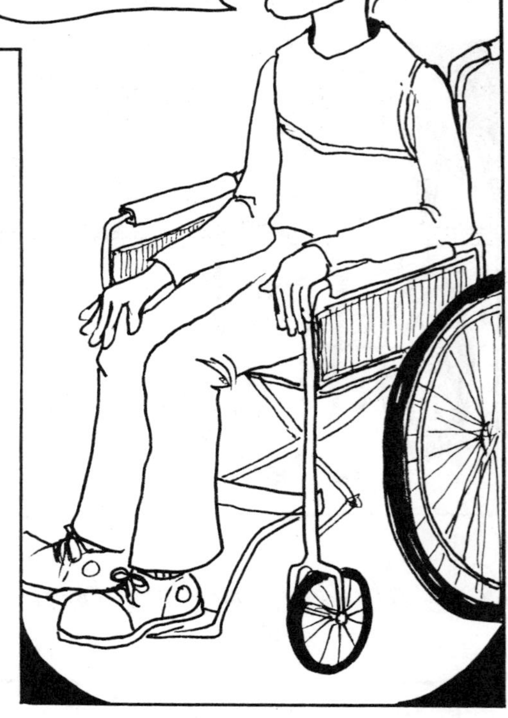

Now let's try . . . **58)2262̄**

58 doesn't go into 2 or 22 but it does go into 226.   This
<u>is about</u> 60 into 230. It could be 3 or 4.   Try 3 first.

$$58\overline{)2262} \quad \frac{3}{}$$
$$\underline{174}$$
$$522$$

Now look at 58 into 522.   This <u>is about</u> 60 into 500.
It could be 8 or 9 but 8 looks closer.   Try 8 first.

$$58\overline{)2262} \quad \frac{38}{}$$
$$\underline{174}$$
$$522$$
$$\underline{464}$$
$$58$$

$$58\overline{)2262} \quad \frac{39}{}$$
$$\underline{174}$$
$$522$$
$$\underline{522}$$
$$0$$

Divide and Conquer 2

★ Now try these. Divide to get the exact answer, using estimation in each step.

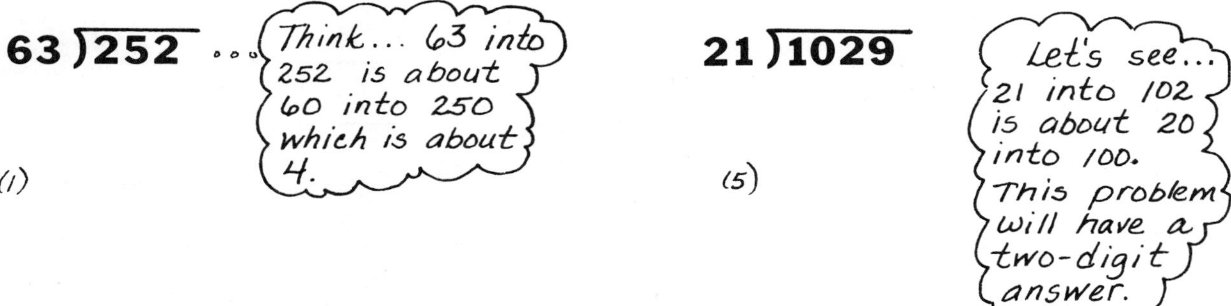

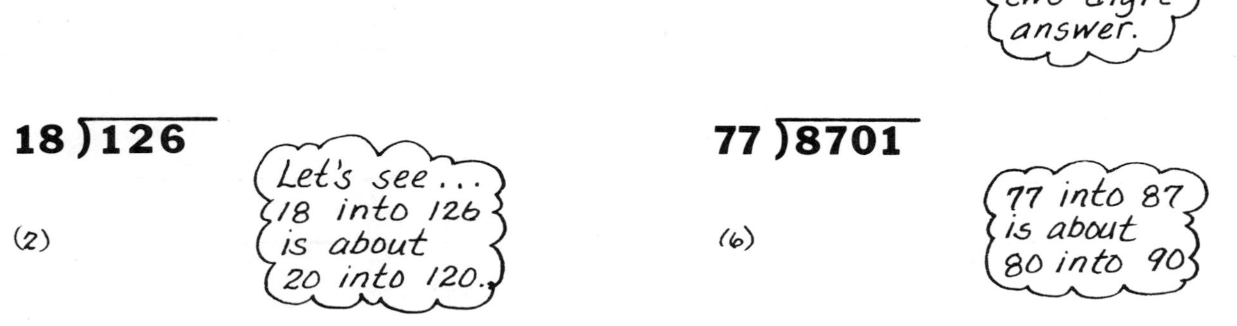

1. 4
2. 7
3. 5
4. 6
5. 49
6. 113
7. 34
8. 325
9.
10.
11.
12.
13.
14.
15.
16.
17.
18.
19.
20.
21.
22.
23.
24.
25.

26.
27.
28.
29.
30.
31.
32.
33.
34.
35.
36.
37.
38.
39.
40.
41.
42.
43.
44.
45.
46.
47.
48.
49.
50.

Divide and Conquer   A

gmp *general math project*

minneapolis public schools

UNIT: EVERYDAY ESTIMATION

DESCRIPTION: In this unit, students will apply their estimation skills to a
variety of real-life problems.

SKILLS TO BE REINFORCED: estimation
adjusting numbers
problem solving

PREREQUISITE SKILLS: estimation

LENGTH: 3 pages;   1-2 days

TEACHER NOTES: You may want to extend this unit by having students think
of additional everyday estimation problems.

NO TEST

EVERYDAY ESTIMATION

Let's see . . .

5% sales tax will be about $10.

So, the total cost of the bike will be about ___________
(1)

I earn $4.15 per hour. I'll have to work about _______
(2)
hours to earn enough for this bike.

Charles has $24 with him. If there were no tax, he could get _______ records. (3) With sales tax, which is about 40¢ per record, he could get _______ records. (4)

Estelle got these running shoes for $21.98.

Benny got the same style shoes on sale for $16.49. He saved about ___________. (5)

This pop is 6/$1.79.

One can will cost about _______ (6)

About how many pro football players could use this elevator at one time?

_______ (7)

Ruth bought a 90 minute blank cassette tape. About how many songs can she record on it? ___________ (8)

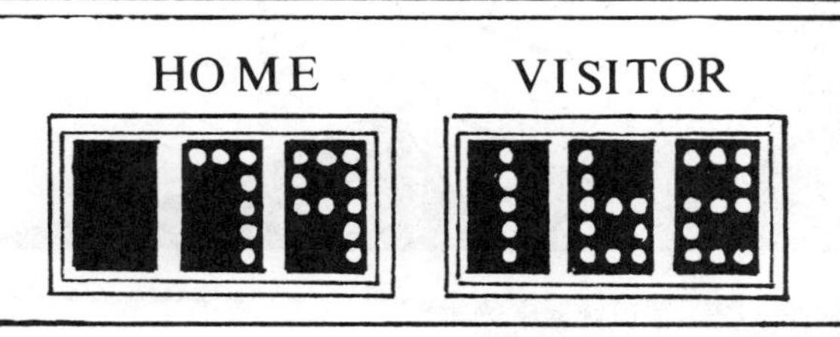

Frank didn't think that the home team's score was correct. Mary had made 22 points; Jane, 19 points; Lisa, 26 points; Felicia, 28 points; and Nanette, 25 points.

According to Frank's estimate, that's about _______ points. (9)

Let's see . . . I'm painting 3 walls of my room. One wall is about 80 sq. ft., another is about 88 sq. ft., and the third is about 64 sq. ft. That's a total of about ______________ sq. ft.
(110)

What's the smallest amount that will cover this area: 1 gallon of paint or 2 or 3 quarts of paint? ______________
(111)

Which costs the least: 1 gallon or 3 quarts? ______________
(112)

This will be the greatest graduation party ever. There are going to be 30 people. Let's see, I've got to make sure I've got enough food.

A pound of hamburger makes 4 or 5 hamburgers. If everyone has 2 "burgers", I'll need about ______________ lbs.
(113)
I'll need about ______________ dozen hamburger buns.
(114)
I'll need about ______________ 1-pound boxes of potato chips.
(115)

The dessert serves about 6-8 people. I'll have to make ______________ desserts to have enough!
(116)

1. $200
2. 50
3. 3
4. 2
5. $5.50
6. $.30
7. 10
8. 30
9. 120
10. 230
11. 3 quarts
12. 1 gallon
13. 12–15
14. 5
15. 3 or 4
16. 5
17.
18.
19.
20.
21.
22.
23.
24.
25.
26.
27.
28.
29.
30.
31.
32.
33.
34.
35.
36.
37.
38.
39.
40.
41.
42.
43.
44.
45.
46.
47.
48.
49.
50.

general math project

minneapolis public schools

UNIT: MAKING YOU PERCENTABLE

DESCRIPTION: In this unit students review some basic percent concepts and then learn to estimate finding the percent of a number. Several uses for percent estimation are then explored.

SKILLS TO BE REINFORCED: estimating with percents

PREREQUISITE SKILLS: adjusting numbers

percent concepts

fraction-percent equivalents

multiplying by $\frac{1}{2}$, $\frac{1}{4}$, $\frac{1}{3}$, $\frac{1}{10}$

LENGTH: 11 pages; 2-3 days

TEACHER NOTES: In the first four pages students estimate with percents using fractions; then a "10%" method of estimating with percents is introduced. To do the problems in the rest of the unit, students may use either fractions or the "10%" method to estimate.

Applications of percent estimation include:
checking exact answers for reasonableness
estimating sales tax
estimating sale prices
estimating bank and loan interest.

NO TEST

MAKING YOU PERCENTABLE !

✱ CIRCLE THE ANSWER

One day, 100% of the 542 students at East High ate school lunch.

How many students ate school lunch?

a) 542

b) 200

c) 0

(1)

On the same day, 0% of the 813 students at Lincoln High ate school lunch.

How many students ate school lunch?

a) 813

b) 368

c) 0

(2)

At Northwest Junior High, 50% of the 444 students ate school lunch.

How many students ate school lunch?

a) 444

b) 222

c) 111

(3)

68% of the 216 students at King School brought a bag lunch.

This is

a) less than half of the students

b) more than half of the students

(4)　　　　　　　　　　　　(4)

<u>About</u> how many students would this be?

a) 52　　　b) 150　　　c) 216

(5)

✱ CIRCLE THE BEST ANSWER

About 23% of all TV programming is news.

That is <u>about</u>

a) $\frac{1}{10}$ b) $\frac{3}{4}$ c) $\frac{1}{4}$

(6)

In 40 hours of programming, <u>about</u> how many hours would be news?

a) 2   b) 10   c) 40

(7)

About 31% of a senior math class are girls.

This is

a) about $\frac{1}{3}$

b) more than $\frac{2}{3}$

(8)

If there are 26 students in the class, <u>about</u> how many are girls?

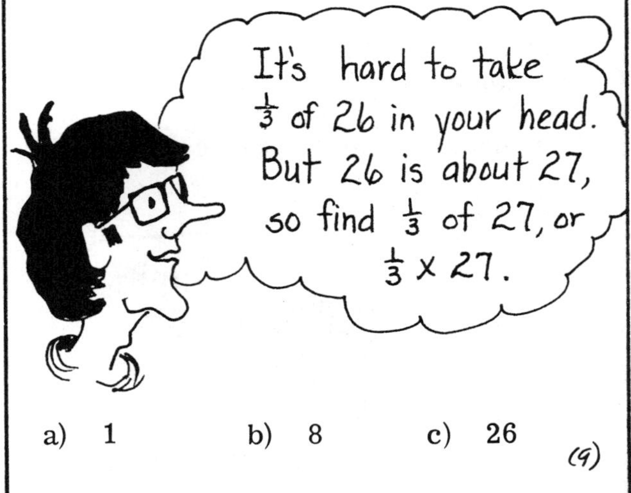

a) 1   b) 8   c) 26

(9)

          Making You Percentable 2

Adjust the numbers so you can work the problems in your head.

9% of the body weight is skin.

That is <u>about</u>

  a) $\dfrac{1}{5}$ b) $\dfrac{1}{10}$ c) $\dfrac{1}{20}$

(10)

For a 123 pound person, the skin weighs <u>about</u>

a) 12 lbs.   b) 60 lbs.   c) 90 lbs.

(11)

About 52% of the babies born in 1974 were boys.

This is <u>about</u>

  a) $\dfrac{1}{3}$ b) $\dfrac{1}{2}$ c) $\dfrac{7}{8}$

(12)

There were 3,159,958 babies born in 1974. <u>About</u> how many were boys?

a) 50,000   b) 1,500,000

c) 3,000,000

(13)

In 1977, there were 444,236,789 radios in use in the United States. 23% of these were in automobiles.

<u>About</u> how many car radios were <u>there</u> in 1977?

a) 13,000,000

b) 111,000,000

c) 444,000,000

(14)

A friend will sell you his bike for 65% of the price he paid for it. He paid $110.

You would pay him <u>about</u>

a) $18

b) $70

c) $100

(15)

AVERAGE
RECOMMENDED
DAILY ALLOWANCE

VITAMIN C	45 mg
VITAMIN A	4500 I.U.
CALCIUM	1200 mg
IRON	18 mg
PROTEIN	51 g

✳ CIRCLE THE BEST ANSWER

If you take a vitamin tablet that gives you 100% of your recommended daily allowance for iron, how many mg of iron will you have taken?

a) 100 mg

b) 18 mg

c) 118 mg

(16)

Sara ate a good breakfast and received about 30% of her recommended daily allowance of protein. This is <u>about</u> how much protein?

a) 15 g

b) 40 g

c) ½ g

(17)

A bowl of cereal claims to provide you with 48% of your daily requirement for Vitamin A. This is <u>about</u>

a) 9000 I.U.

b) 4800 I.U.

c) 2200 I.U.

(18)

You eat enough so that you have had 40 mg of Vitamin C. <u>About</u> what percent of the recommended daily allowance have you had?

a) 90%

b) 40%

c) 20%

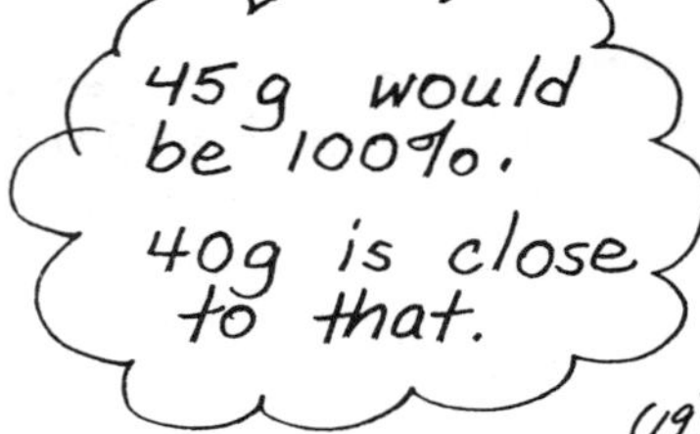

(19)

In an average day, Fred takes in about 650 mg of calcium. That is <u>about</u> what percent of the recommended <u>daily</u> allowance?

a) 95%   b) 50%   c) 15%

(20)

63% OF **48** IS ABOUT ________ **?**

1. ADJUST ONE OR BOTH NUMBERS TO GET AN ADJUSTED PROBLEM.

Adjust the percent to the nearest 10%:

63% ⟶ 60%

Adjust the number.

48 ⟶ 50

63% OF **48** IS ABOUT **60%** OF **50**

2. COMPUTE THE ADJUSTED PROBLEM (60% of 50).

60% x 50   is the same as   6 x (10% x 50)

Find 10% x 50: ________ (21)

Multiply that by 6: ________ (22)

SO **63%** OF **48** IS ABOUT **30** .

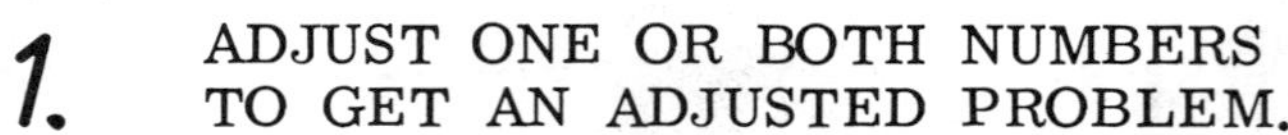

O.K. I'll try!

1. ADJUST ONE OR BOTH NUMBERS TO GET AN ADJUSTED PROBLEM.

37% ⟶ 40%

20 can be left as it is.

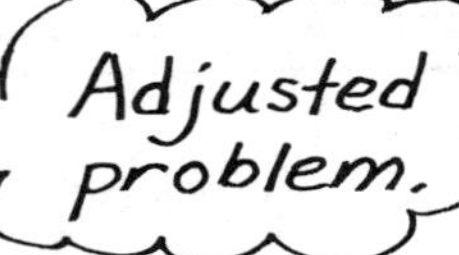

37% OF 20 IS ABOUT 40% OF 20.

2. COMPUTE THE ADJUSTED PROBLEM (40% of 20).

Take 10% x 20: __________ (23)

Multiply that by 4: __________ (24)

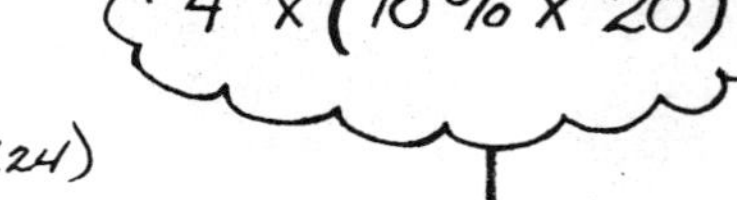

SO **37% OF 20** IS ABOUT __________. (25)

1. 28% of 87 is about __________ x __________ = __________.
(26)   (27)   (28)

2. 52% of $18.45 is about __________ x __________ = $ __________.
(29)   (30)   (31)

3. 18% of 296 is about __________ x __________ = __________.
(32)   (33)   (34)

4. 17% of Minnesota's population of 3,975,069 people live in Minneapolis and St. Paul. About how many people live in the Twin Cities?

17% of 3,975,069 is about ______ x __________ = __________.
(35)   (36)   (37)

5. 72% of the states in the United States require students to attend school until the age of 16. About how many states require this? __________

HOOKED ON ESTIMATING

To find out what this sign says, draw a line from each problem on the left to the closest estimate on the right. Each line will go through a letter and a number. Put the letter above the number in the sign. The first one is done for you.

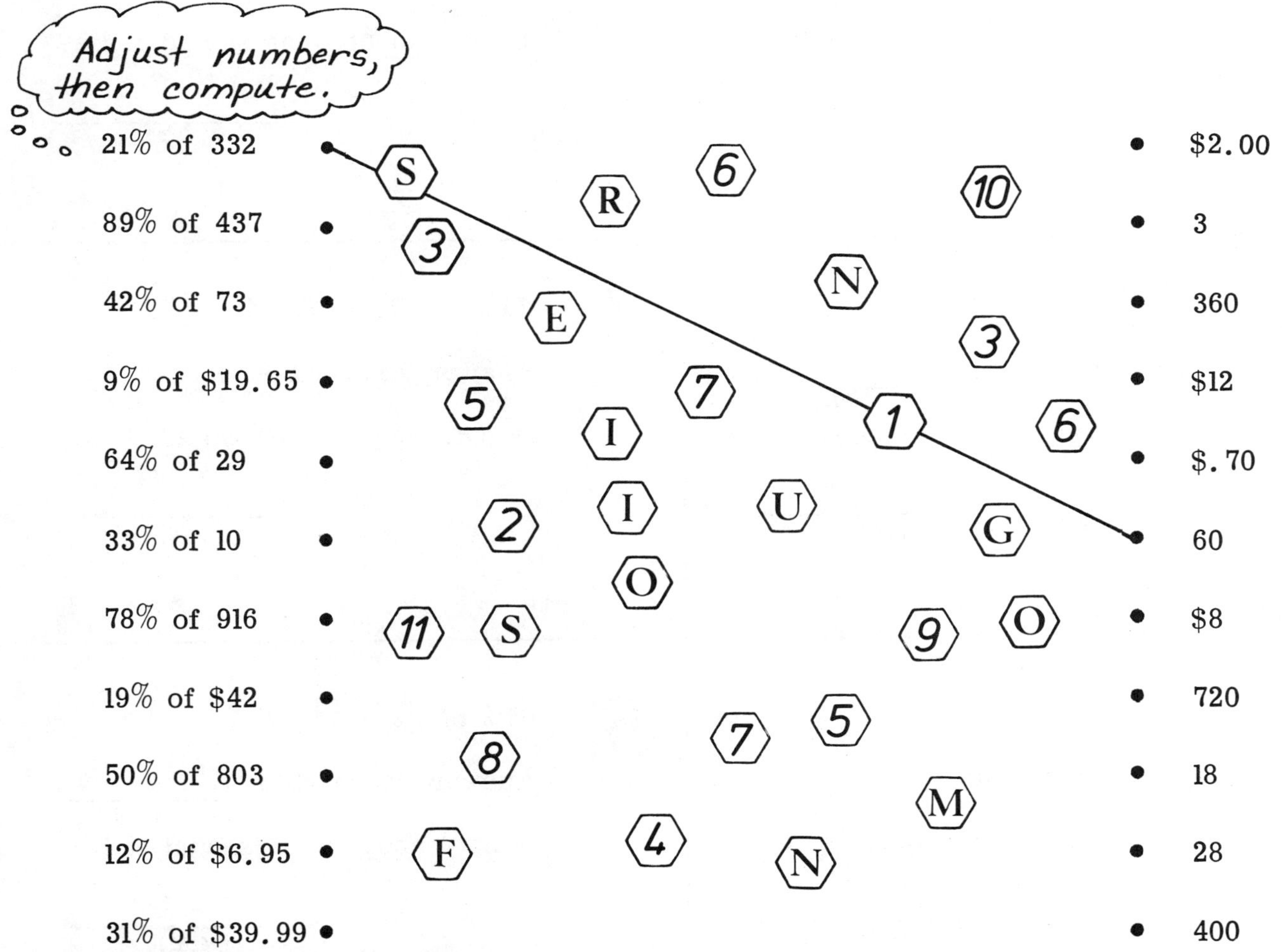

✻ Use ESTIMATION to tell you whether the exact answer to each problem is reasonable. Use fractions or the "10%" method to estimate.

1. 11% of 92 = 1.102

Answer reasonable? ______
(39)

If not, <u>about</u> what should it be?

______
(40)

2. 29% of 158 = 45.82

Answer reasonable? ______
(41)

If not, <u>about</u> what should it be?

______
(42)

3. 84% of 78 = 655.2

Answer reasonable? ______
(43)

If not, <u>about</u> what should it be?

______
(44)

4. 36% of 374 = 97.24

Answer reasonable? ______
(45)

If not, <u>about</u> what should it be?

______
(46)

5. 51% of 218 = 111.18

Answer reasonable? ______
(47)

If not, <u>about</u> what should it be?

______
(48)

6. 67% of 22 = 14.74

Answer reasonable? ______
(49)

If not, <u>about</u> what should it be?

______
(50)

✱ Using the price information above, ESTIMATE the sale price.

1. 

On sale, these shorts are 23% off the regular price of $9.99. <u>About</u> how much are the <u>shorts</u>?

23% of $9.99 is about

___________ x ___________ =
  (51)   (52)

___________ .
   (53)

Subtract that from $10.00 ∘∘
the adjusted regular price.

So, you'd pay <u>about</u>

___________ for the shorts.
  (54)

2. <u>About</u> how much would a soccer ball cost?

3. <u>About</u> how much would a pair of soccer shoes cost?

Look at this example:

Carlos has $200 to spend. Can he buy the stereo if the total price includes sales tax?

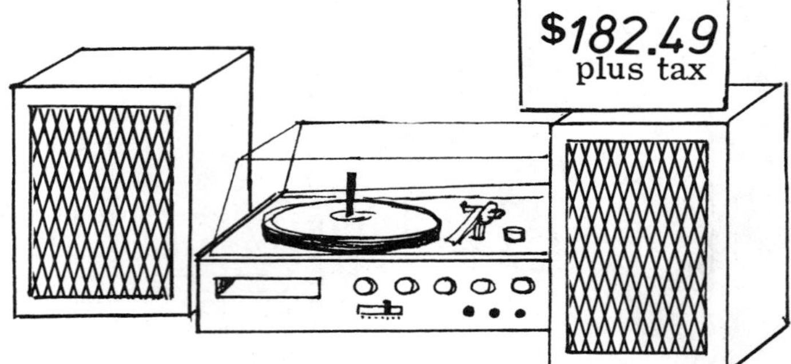

TO ESTIMATE A 5% SALES TAX ON AN ITEM ...

1. ADJUST THE COST OF THE ITEM TO GET AN ADJUSTED PROBLEM.

5% of $182.49 is <u>about</u> 5% of $180.

2. COMPUTE THE ADJUSTED PROBLEM (5% of $180).

5% x $180 is the same as $\frac{1}{2}$ x (10% x $180).

Find 10% x $180: _______ (57)

Take $\frac{1}{2}$ of that: _______ (58)

Does Carlos have enough money to buy the stereo? _______ (59)

ESTIMATE the tax on these items :

A ten-speed bike costs $129.99. <u>About</u> how much is the tax?

_______ (60)

A new motorcycle costs $899.50. <u>Estimate</u> the tax.

_______ (61)

You are going out to eat and the bill is $17.83. <u>About</u> how much is the tax?

_______ (62)

Your aunt buys a new car that costs $5987. What is the estimated tax on the car?

_______ (64)

You have $18. Do you have enough money? _______ (63)

Will $6300 to buy the car?

_______ (65)

A major credit card company charges 22% interest on any unpaid balance. If you have an unpaid balance of $360, about how much finance charge will you have to pay?

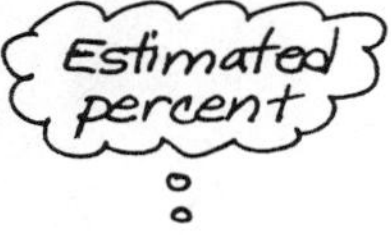

$$\underline{\qquad}\% \times \$360 = \underline{\qquad}$$
(66)   (67)

Jim and Sue plan to start a lawn care service and need to buy a lawn mower. They have some cash, but want to borrow $200. Bank interest for the loan is 16% per year.

About how much interest will they pay in a year?

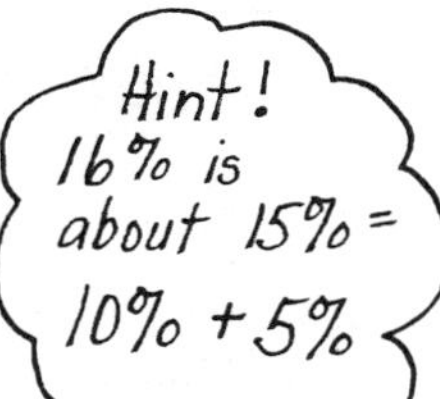

$$\underline{\qquad}$$
(68)

Mr. Payne has $5000 and wants to put it in the bank. He can choose a passbook account that will pay him $5\frac{1}{2}\%$ interest per year or a savings certificate that will pay 11% interest per year.

About how much interest will he get on the $5\frac{1}{2}\%$ account in a year?

$$\underline{\qquad}$$
(69)

About how much interest will he get on the 11% account in a year?

$$\underline{\qquad}$$
(70)

About how much more will he earn with the savings certificate?

$$\underline{\qquad}$$
(71)

Jerry is going to buy a car. The car costs $2195.

About how much will he have to pay when he figures 5% sales tax and a 17% bank loan.

$\underline{\qquad}$ +	$\underline{\qquad}$ +	$\underline{\qquad}$ =	$\underline{\qquad}$
(72)	(73)	(74)	(75)
est. car cost	est. tax	est. loan interest	est. total

1. a) 542
2. c) 0
3. b) 222
4. b) more than half of the students
5. b) 150
6. c) $\frac{1}{4}$
7. b) 10
8. a) about $\frac{1}{3}$
9. b) 8
10. b) $\frac{1}{10}$
11. a) 12 lbs.
12. b) $\frac{1}{2}$
13. b) 1,500,000
14. b) 111,000,000
15. b) $ 70
16. b) 18 mg
17. a) 15 g
18. c) 2200 I.U.
19. a) 90%
20. b) 50%
21. 5
22. 30
23. 2
24. 8
25. 8

26. 30%
27. 90
28. 27
29. 50%
30. 20
31. $ 10
32. 20%
33. 300
34. 60
35. 20%
36. 4,000,000
37. 800,000
38. 35
39. no
40. 9
41. yes
42. —
43. no
44. 64
45. no
46. 160
47. yes
48. —
49. yes
50. —

51. 20%
52. $10.00
53. $2.00
54. $8.00
55. $17.00
56. $9 or $10
57. $18
58. $9
59. yes
60. $6.50
61. $45
62. $.90
63. no
64. $300
65. yes
66. 20
67. $72
68. $30
69. $250
70. $500
71. $250
72. 2000
73. 100
74. 400
75. $2500

PUZZLE ANSWER

page 6   (Message Puzzle)

GONE FISSION

Depending on whether students estimate using fractions or the "10%" method, their answers may differ slightly from those given. From #21 on, Answer Key answers are all based on the "10%" method.

general math project

minneapolis public schools

UNIT: FRACTION ACTION

DESCRIPTION: In this unit, students will extend their estimating skills to working with fractions. They will learn to adjust fractional numbers to ones that can be easily worked with in order to determine if answers are reasonable.

SKILLS TO BE REINFORCED: fractional relationships

estimation

PREREQUISITE SKILLS: addition, subtraction, multiplication and division of fractions

LENGTH: 11 pages;   2-3 days

TEACHER NOTES: Fraction estimation is a difficult concept for most students. It should be stressed and reinforced throughout the year.

We've limited the fractions used to halves, thirds, fourths, sixths, eighths and tenths.

NO TEST

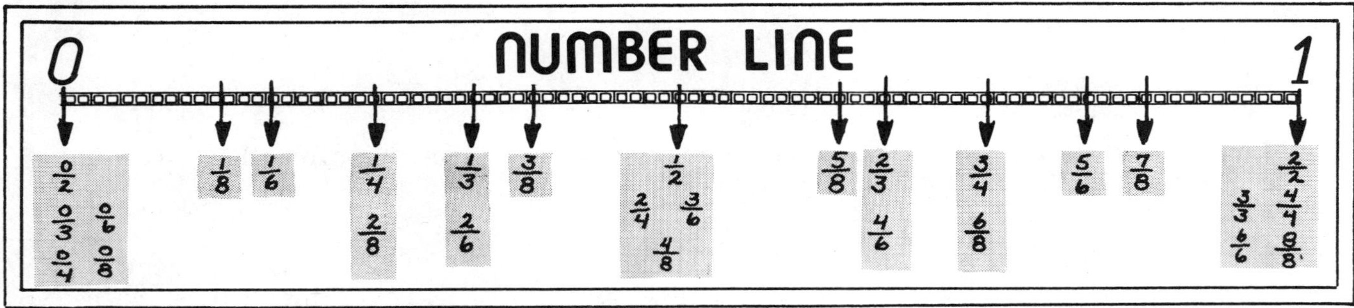

List the equivalent fractions to these fractions.   Use the number line.

$$\frac{2}{3} = \underline{\hspace{4cm}}$$ (1)

$$\frac{1}{3} = \underline{\hspace{4cm}}$$ (3)

$$\frac{1}{4} = \underline{\hspace{4cm}}$$ (2)

$$1 = \underline{\hspace{4cm}}$$ (4)

Tell which fraction is bigger by using the number line on page 1.

$\dfrac{3}{8}$ or $\dfrac{2}{3}$ _________ (5) $\dfrac{2}{6}$ or $\dfrac{1}{8}$ _________ (7)

$\dfrac{5}{6}$ or $\dfrac{3}{4}$ _________ (6) $\dfrac{5}{8}$ or $\dfrac{4}{6}$ _________ (8)

Here's how we can show that one number is larger than, smaller than or equal to another number.

$>$ means "is greater than"; $5 > 4$

$=$ means "is equal to" ; $5 = 5$

$<$ means "is less than" ; $4 < 5$

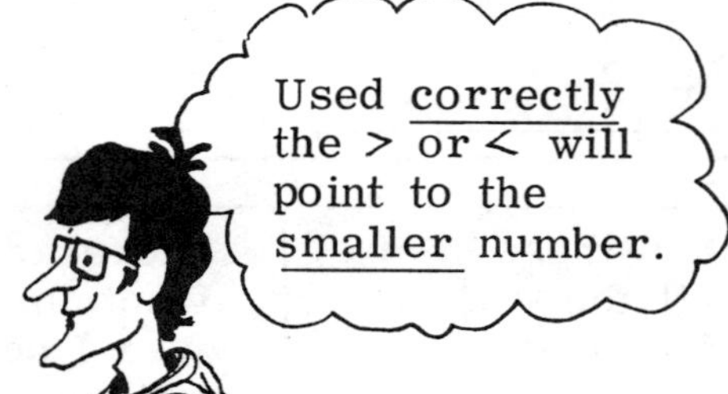

Circle T or F to tell whether each statement is correct or not. Use the number line on page 1 if you need to.

$\dfrac{1}{2} > \dfrac{1}{3}$ **T F** (9)	$\dfrac{1}{3} > \dfrac{1}{6}$ **T F** (10)
$\dfrac{1}{4} = \dfrac{2}{6}$ **T F** (11)	$\dfrac{2}{3} > \dfrac{5}{8}$ **T F** (12)
$\dfrac{1}{4} < \dfrac{2}{8}$ **T F** (13)	$\dfrac{5}{6} < \dfrac{3}{4}$ **T F** (14)

Put in the correct symbol: $>$, $=$, or $<$.
Use the number line if you need to.

$\dfrac{5}{6}$ ___ $\dfrac{1}{8}$ (15)	$\dfrac{3}{4}$ ___ $\dfrac{1}{6}$ (16)
$\dfrac{1}{8}$ ___ $\dfrac{5}{8}$ (17)	$\dfrac{1}{6}$ ___ $\dfrac{1}{4}$ (18)
$\dfrac{2}{4}$ ___ $\dfrac{3}{6}$ (19)	$\dfrac{5}{6}$ ___ $\dfrac{1}{2}$ (20)

To __estimate__ the answers to problems involving fractions...

① Adjust the fractions to get numbers easier to work with, but close to the exact fractions.

② Work the adjusted problem!

Using the steps above, look at this addition example:

$$\frac{3}{4} + \frac{7}{8} =$$

① $\frac{3}{4}$ is close to 1 $\frac{7}{8}$ is close to 1

So $\frac{3}{4} + \frac{7}{8}$ is __about__ 1 + 1 ... *This is the adjusted problem.*

② 1 + 1 = 2, so $\frac{3}{4} + \frac{7}{8}$ should be __about__ 2.

Circle the best adjusted number for each given fraction. Look at the number line on page 1 for help.

$\frac{3}{8}$	0	$\frac{1}{2}$	1	(21)	$\frac{5}{8}$	0	$\frac{1}{2}$	1	(24)
$\frac{5}{6}$	0	$\frac{1}{2}$	1	(22)	$\frac{1}{3}$	0	$\frac{1}{2}$	1	(25)
$\frac{1}{6}$	0	$\frac{1}{2}$	1	(23)	$\frac{2}{3}$	0	$\frac{1}{2}$	1	(26)

     Fraction Action   3

✱ Circle the best adjusted number for each given mixed number.

$9\frac{1}{6}$	$8\frac{1}{2}$	9	$9\frac{1}{2}$ (27)	$1\frac{3}{4}$	1	2	$2\frac{1}{2}$ (30)
$3\frac{7}{8}$	3	$3\frac{1}{2}$	4 (28)	$10\frac{3}{8}$	10	$10\frac{1}{2}$	$2\frac{1}{2}$ (31)
$16\frac{5}{8}$	16	17	18 (29)	$6\frac{1}{4}$	6	7	$7\frac{1}{2}$ (32)

✱ Circle the best estimated answer for each problem.

$7 + 1\frac{1}{4}$:	8	$9\frac{1}{2}$	$10\frac{1}{2}$ (33)	$\frac{8}{10} + \frac{7}{8}$:	1	$1\frac{1}{3}$	2 (37)
$\frac{3}{8} + \frac{1}{6}$:	$\frac{1}{8}$	$\frac{1}{2}$	1 (34)	$9\frac{7}{8} + 2\frac{3}{4}$:	12	13	14 (38)
$\frac{1}{3} + \frac{1}{2}$:	$\frac{3}{8}$	1	$1\frac{1}{2}$ (35)	$17\frac{1}{4} + 8\frac{8}{10}$:	24	25	26 (39)
$\frac{3}{4} + \frac{2}{3}$:	$\frac{7}{8}$	1	$1\frac{1}{2}$ (36)	$6\frac{1}{8} + 5\frac{2}{3}$:	10	$11\frac{1}{2}$	13 (40)

Try these !

Jessie had a pattern that took 5/8 yard for a vest and $1\frac{1}{4}$ yards for slacks. He found a material remnant 2 yards long that was on sale. Will it be enough for the vest <u>and</u> slacks?

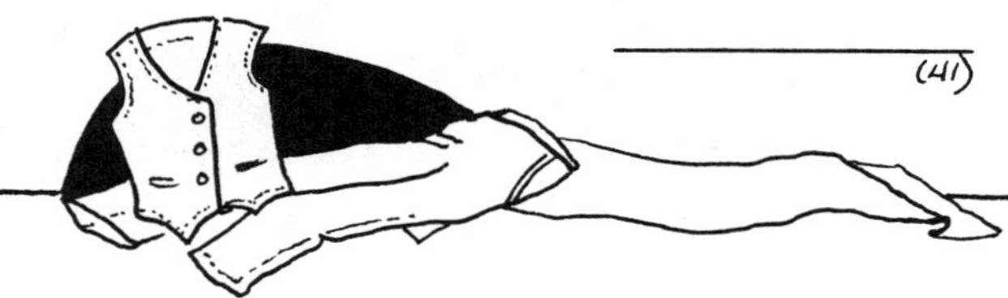

_________________ (41)

A toy manufacturer makes a toy car. The cost of making the parts is:

body:	$3\frac{5}{8}$ ¢
wheels:	$\frac{5}{8}$ ¢
frame:	$2\frac{1}{4}$ ¢
steering wheel:	$1\frac{1}{4}$ ¢

About how much does it cost to make the car? _________________ (42)

Jackie is making a brick and board bookcase. It has 2 shelves, each $3\frac{3}{8}$ feet long. Estimate the amount of wood needed for the shelves.

_________________ (43)

She has found two boards that she thinks will be long enough. One is 6 feet long and the other is 8 feet long. Which should she choose to have enough wood? _________________ (44)

Adjust the fractions in each subtraction problem and then circle the best estimated answer to the problem.

$10\frac{2}{3} - 5$ 4   6   8   (45)

$16 - 5\frac{1}{3}$ 10   11   12   (48)

$9\frac{1}{2} - 3\frac{3}{4}$ $3\frac{1}{2}$ $5\frac{1}{2}$ 7   (46)

$17\frac{1}{4} - 9\frac{5}{6}$ 7 $8\frac{1}{2}$ 9   (49)

$6\frac{3}{8} - 2\frac{1}{3}$ 4   3 $3\frac{1}{2}$ (47)

$9\frac{1}{6} - 2\frac{1}{3}$ 6 $6\frac{1}{2}$ 8   (50)

Kim did this worksheet. Without working the problems, <u>mentally</u> check Kim's work. Tell whether or not each of her answers is reasonable.

Math Worksheet 7		NAME _Kim B._
⭐1 $\begin{array}{r} 8\frac{1}{2} \\ 2\frac{1}{4} \\ \frac{1}{2} \\ +\ \frac{1}{8} \\ \hline 10\frac{3}{8} \end{array}$	⭐2 $\begin{array}{r} 18\frac{5}{6} \\ -12\frac{1}{4} \\ \hline 6\frac{3}{4} \end{array}$	⭐3 $\begin{array}{r} 19\frac{1}{8} \\ -6\frac{3}{4} \\ \hline 13\frac{5}{8} \end{array}$
Reasonable? _______ (51)	Reasonable? _______ (52)	Reasonable? _______ (53)
⭐4 $\begin{array}{r} 9\frac{1}{4} \\ 2\frac{1}{3} \\ +\ \frac{3}{4} \\ \hline 12\frac{1}{3} \end{array}$	⭐5 $\begin{array}{r} 16\frac{1}{8} \\ -11\frac{7}{8} \\ \hline 5\frac{6}{8} \end{array}$	⭐6 $\begin{array}{r} 16\frac{1}{2} \\ +5\frac{1}{8} \\ \hline 25\frac{3}{8} \end{array}$
Reasonable? _______ (54)	Reasonable? _______ (55)	Reasonable? _______ (56)

    Fraction Action 5

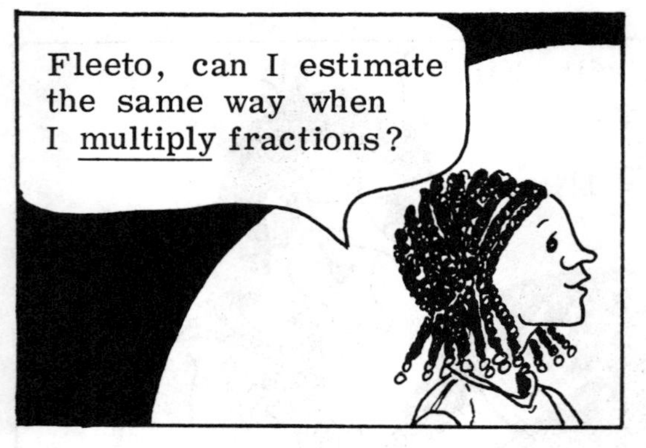

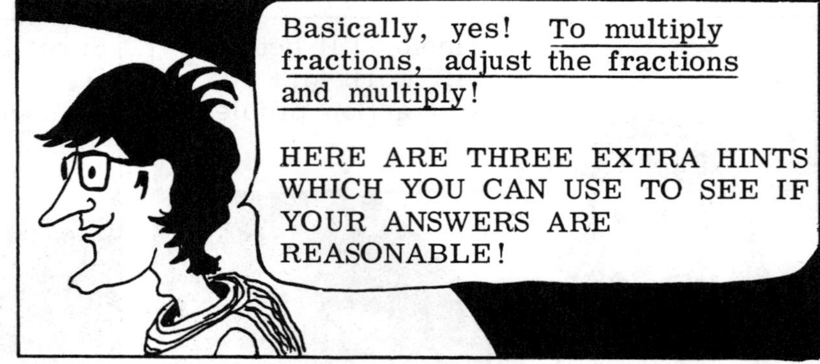

1 WHEN YOU MULTIPLY TWO PROPER FRACTIONS, YOUR ANSWER WILL ALWAYS BE LESS THAN 1!

These are all proper fractions - in each, the numerator is smaller than the denominator.

$$\frac{1}{4} \times \frac{1}{2} = \frac{1}{8}$$
$$\frac{2}{3} \times \frac{3}{4} = \frac{6}{12} = \frac{1}{2}$$
$$\frac{5}{6} \times \frac{1}{8} = \frac{5}{48}$$

These answers are all less than 1 !

2 WHEN YOU MULTIPLY A WHOLE OR MIXED NUMBER AND A PROPER FRACTION, YOUR ANSWER WILL ALWAYS BE LESS THAN THE WHOLE OR MIXED NUMBER!

Each has a whole or mixed number times a proper fraction.

$$50 \times \frac{1}{2} = 25$$
$$27 \times \frac{2}{3} = 18$$
$$7\frac{1}{2} \times \frac{1}{4} = 1\frac{7}{8}$$

Each answer is less than the whole or mixed number.

✱ Now try these!

$41 \times \frac{1}{4}$	The answer will be a) more than 41 b) less than 41 (57)	The answer is <u>about</u> a) 10 b) 30 c) 50 (58)	$41 \times \frac{1}{4}$ is about $40 \times \frac{1}{4}$
$\frac{7}{8} \times \frac{1}{3}$	The answer will be a) more than 1 b) less than 1 (59)	The answer is <u>about</u> a) $\frac{1}{3}$ b) $\frac{5}{6}$ c) $1\frac{2}{3}$ (60)	$\frac{7}{8} \times \frac{1}{3}$ is about $1 \times \frac{1}{3}$
$31 \times \frac{2}{3}$	The answer will be a) more than 31 b) less than 31 (61)	The answer is <u>about</u> a) 11 b) 20 c) 40 (62)	$31 \times \frac{2}{3}$ is about $30 \times \frac{2}{3}$

3 WHEN YOU MULTIPLY TWO MIXED NUMBERS, YOUR ANSWER WILL BE LARGER THAN EITHER OF THE MIXED NUMBERS.

Each has two mixed numbers multiplied together.

$$2\frac{1}{4} \times 4\frac{7}{8} = 10\frac{31}{32}$$
$$4\frac{1}{3} \times 6\frac{1}{3} = 27\frac{4}{9}$$

Each answer is larger than either of the two mixed numbers.

✱ Adjust each mixed number and then circle the best estimated answer.

$1\frac{3}{4} \times 10\frac{1}{8}$ is about: a) 10 b) 20 c) 30 (63)

$3\frac{2}{3} \times 7\frac{1}{6}$ is about: a) 10 b) 21 c) 28 (64)

These are fighting words!

$$\frac{}{6} \ \frac{}{8} \ \frac{}{3} \ \frac{}{7} \ \frac{T}{1} \quad \frac{}{5} \quad \frac{}{10} \ \frac{}{9} \ \frac{}{2} \ \frac{}{4}$$

To answer this question, work each problem below. Find each answer in the Code Box and put the corresponding letter above the problem number. The first one is done for you.

Circle the best estimated answer for each problem.

1. $\frac{7}{8} \times \frac{7}{8}$ is about: $\frac{1}{4}$ ① 2

2. $31 \times \frac{1}{8}$ is about: 2 4 6

3. $2\frac{1}{8} \times 6\frac{2}{3}$ is about: 9 11 14

4. $\frac{1}{2} \times \frac{3}{5}$ is about: $\frac{1}{4}$ 1 3

5. $8 \times \frac{2}{3}$ is about: 3 6 8

6. $\frac{1}{3} \times 22$ is about: 2 $5\frac{1}{3}$ 7

7. $1\frac{1}{6} \times 1\frac{3}{4}$ is about: 2 $4\frac{1}{2}$ $6\frac{1}{2}$

8. $5\frac{2}{3} \times 6\frac{1}{4}$ is about: 12 36 40

9. $\frac{1}{2} \times 1\frac{1}{8}$ is about: $\frac{1}{8}$ $\frac{1}{2}$ $\frac{3}{4}$

10. $19\frac{7}{8} \times 3\frac{3}{4}$ is about: 57 60 80

Code Box

T	O	T	U	U	A	A	O	B	B
$\frac{1}{4}$	$\frac{1}{2}$	1	2	4	6	7	14	36	80

Fraction Action 7

Look at these examples...

✱ Estimate the answer to $\frac{3}{4} \div \frac{7}{8}$

$\frac{3}{4} \div \frac{7}{8}$ is about $1 \div 1 = 1$

To be reasonable, the answer to $\frac{3}{4} \div \frac{7}{8}$ should be close to 1.

✱ Estimate the answer to $10\frac{1}{3} \div 20\frac{2}{3}$

$10\frac{1}{3} \div 20\frac{2}{3}$ is about $10 \div 21$ which is close to $10 \div 20 = \frac{1}{2}$.

To be reasonable, the answer to $10\frac{1}{3} \div 20\frac{2}{3}$ should be about $\frac{1}{2}$.

✱ Estimate: $8\frac{5}{6} \div 2\frac{3}{4}$

$8\frac{5}{6} \div 2\frac{3}{4}$ is about $9 \div \underline{} = \underline{}$
 (65) (66)

To be reasonable, the answer to $8\frac{5}{6} \div 2\frac{3}{4}$ should be about $\underline{}$.
 (67)

✱ Estimate: $2\frac{1}{6} \div 7\frac{3}{4}$

$2\frac{1}{6} \div 7\frac{3}{4}$ is about $\underline{} \div 8 = \underline{}$
 (68) (69)

To be reasonable, the answer to $2\frac{1}{2} \div 7\frac{3}{4}$ should be about $\underline{}$.
 (70)

$10\frac{1}{4} \div \frac{1}{4}$ is about $10 \div \frac{1}{4} = 10 \times 4 = 40$.

The answer to $10\frac{1}{4} \div \frac{1}{4}$ should be about 40!

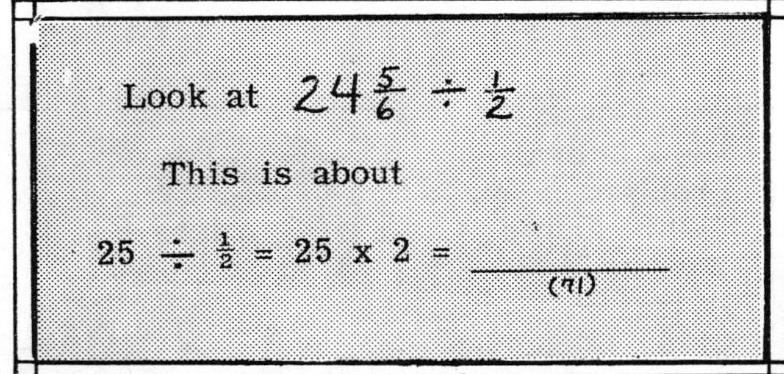

The answer to $24\frac{5}{6} \div \frac{1}{2}$ should be about __________ . (72)

Estimate the answer to each problem.

* $8\frac{1}{4} \div \frac{1}{3}$ is about __________ ÷ __________ = __________ (73)

* $6\frac{3}{8} \div 2\frac{1}{3}$ is about __________ ÷ __________ = __________ (74)

* $9\frac{5}{6} \div \frac{1}{6}$ is about __________ ÷ __________ = __________ (75)

* $3\frac{3}{4} \div \frac{7}{8}$ is about __________ ÷ __________ = __________ (76)

* $1\frac{1}{3} \div 4\frac{1}{6}$ is about __________ ÷ __________ = __________ (77)

1 EVERY TIME YOU DIVIDE A NUMBER BY A LARGER NUMBER, THE ANSWER SHOULD BE LESS THAN 1.

2 EVERY TIME YOU DIVIDE A NUMBER BY A SMALLER NUMBER, THE ANSWER SHOULD BE MORE THAN 1.

3 WHEN YOU DIVIDE A NUMBER BY A NUMBER ABOUT THE SAME SIZE, THE ANSWER IS CLOSE TO 1.

Here are two examples.

Rachel did this problem in class. Is her answer reasonable?

To answer this question, think . . .

Since 7 is divided by a smaller number, 6, the answer should be more than 1.

In addition, 7 and 6 are <u>about</u> the same size, so the answer should also be close to 1.

Estimate: $7\frac{1}{3} \div 6\frac{1}{3}$ is about $7 \div 6 = \frac{7}{6}$

Is $3\frac{8}{9}$ a reasonable answer? ______
(78)

Dale did this problem in class. Is his answer reasonable?

Should the answer be less than **1** or more than 1? ______________
(79)

Should the answer also be fairly close to 1? ________
(80)

Estimate the answer:
$5\frac{3}{4} \div 1\frac{3}{4}$ is about

______ ÷ ______ = ______
(81)   (82)   (83)

Is $3\frac{2}{7}$ a reasonable answer? ______
(84)

     Fraction Action   10

$$\frac{7}{8} \div \frac{1}{4} = 3\frac{1}{2}$$

Should the answer be more or less than 1? _______
(85)

Should the answer also be fairly close to 1? _______
(86)

Estimate the answer:

$\frac{7}{8} \div \frac{1}{4}$ is about

Are $\frac{7}{8}$ and $\frac{1}{4}$ about the same size?

$\frac{}{(87)} \div \frac{1}{4} = \frac{}{(88)}$

Is $3\frac{1}{2}$ a reasonable answer? _______
(89)

$$1\frac{3}{4} \div 7\frac{2}{3} = 4\frac{1}{8}$$

Should the answer be more or less than 1? _______
(90)

Should it also be fairly close to 1? _______
(91)

Estimate the answer:

$1\frac{3}{4} \div 7\frac{2}{3}$ is about

$\frac{}{(92)} \div \frac{}{(93)} = \frac{}{(94)}$

Is $4\frac{1}{8}$ a reasonable answer? _______
(95)

$$\frac{3}{8} \div \frac{1}{2} = \frac{3}{4}$$

Should the answer be more or less than 1? _______
(96)

Should it also be fairly close to 1? _______
(97)

Estimate the answer:

$\frac{3}{8} \div \frac{1}{2}$ is about

$\frac{3}{8}$ is about $\frac{1}{2}$

$\frac{1}{2} \div \frac{}{(98)} = \frac{}{(99)}$

Is $\frac{3}{4}$ a reasonable answer? _______
(100)

     Fraction Action   11

1. $\frac{4}{6}$

2. $\frac{2}{8}$

3. $\frac{2}{6}$

4. $\frac{2}{2}, \frac{3}{3}, \frac{4}{4}, \frac{5}{5}, \frac{6}{6}$

5. $\frac{2}{3}$

6. $\frac{5}{6}$

7. $\frac{2}{6}$

8. $\frac{4}{6}$

9. T

10. T

11. F

12. T

13. F

14. F

15. <

16. >

17. >

18. <

19. =

20. >

21. $\frac{1}{2}$

22. 1

23. 0

24. $\frac{1}{2}$

25. $\frac{1}{2}$

26. $\frac{1}{2}$

27. 9

28. 4

29. 17

30. 2

31. $10\frac{1}{2}$

32. 6

33. 8

34. $\frac{1}{2}$

35. 1

36. $1\frac{1}{2}$

37. 2

38. 13

39. 26

40. $11\frac{1}{2}$

41. yes

42. $7\frac{1}{2}$ ¢ or 8 ¢

43. 7 feet

44. 8 feet

45. 6

46. $5\frac{1}{2}$

47. 4

48. 11

49. 7

50. $6\frac{1}{2}$

#		#	
51.	no	76.	$4 \div 1 = 4$
52.	yes	77.	$1 \div 4 = \frac{1}{4}$
53.	no	78.	no
54.	yes	79.	more than 1
55.	no	80.	no
56.	no	81.	6
57.	b) less than 41	82.	2
58.	a) 10	83.	3
59.	b) less than 1	84.	yes
60.	a) $\frac{1}{3}$	85.	more than 1
61.	b) less than 31	86.	no
62.	b) 20	87.	1
63.	b) 20	88.	4
64.	c) 28	89.	yes
65.	3	90.	less than 1
66.	3	91.	no
67.	3	92.	2
68.	2	93.	8
69.	$\frac{1}{4}$	94.	$\frac{1}{4}$
70.	$\frac{1}{4}$	95.	no
71.	50	96.	less than 1
72.	50	97.	yes
73.	$8 \div \frac{1}{3} = 24$	98.	$\frac{1}{2}$
74.	$6 \div 2 = 3$	99.	1
75.	$10 \div \frac{1}{6} = 60$	100.	yes
		101.	103

PUZZLE ANSWER
page 7 (Message Puzzle):
ABOUT A BOUT

general math project

minneapolis public schools

UNIT: THE MONEY ROUND-UP

DESCRIPTION: In this unit, students learn a technique for overestimating when they must make sure they have enough of something. They do this by rounding numbers up rather than just rounding them off.

SKILLS TO BE REINFORCED: rounding numbers up

estimation

PREREQUISITE SKILLS: adjusting numbers by rounding

LENGTH: 5 pages;   1-2 days

TEACHER NOTES: This technique of rounding-up to make sure there is enough of something extends the estimation ideas already covered.

NO TEST

THE MONEY ROUND-UP

They each use a different method of estimating to see if $1.00 is enough.

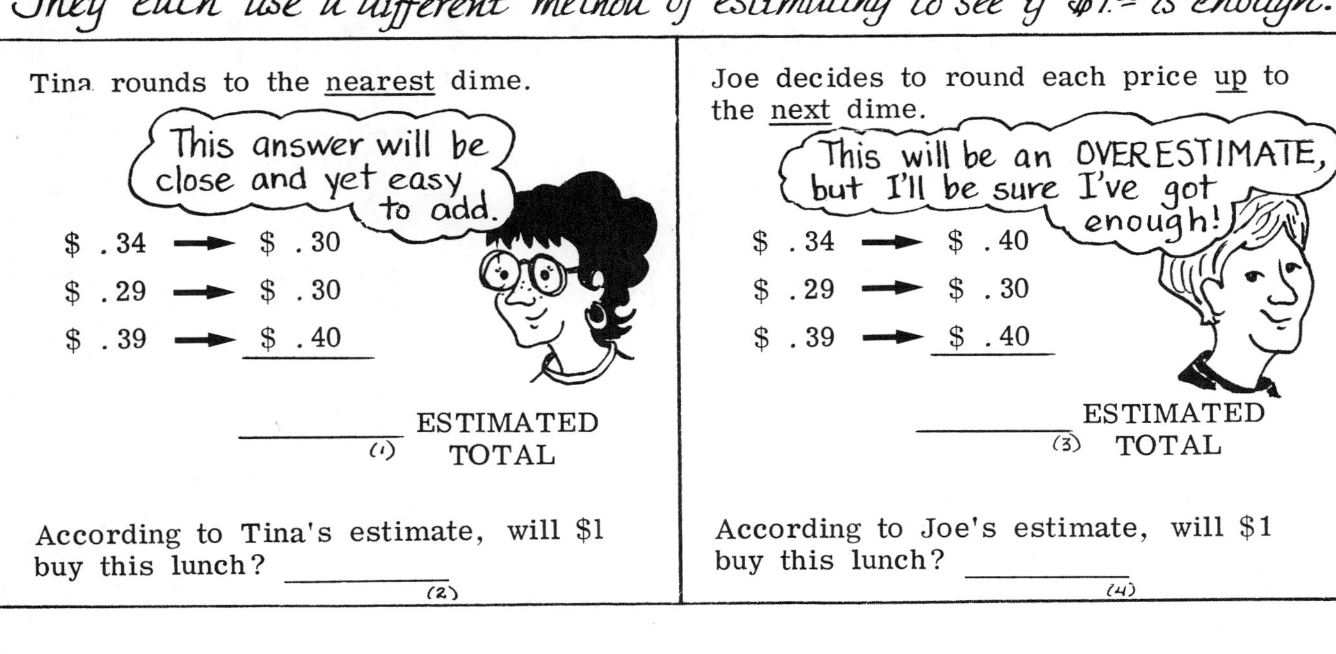

Tina rounds to the <u>nearest</u> dime.

$ .34 → $ .30
$ .29 → $ .30
$ .39 → $ .40

__________ ESTIMATED TOTAL
(1)

According to Tina's estimate, will $1 buy this lunch? __________
(2)

Joe decides to round each price <u>up</u> to the <u>next</u> dime.

$ .34 → $ .40
$ .29 → $ .30
$ .39 → $ .40

__________ ESTIMATED TOTAL
(3)

According to Joe's estimate, will $1 buy this lunch? __________
(4)

What is the <u>exact</u> total for a BBQ, fries and pop? __________
(5)

Tina's estimate was very close to the exact cost. Why wasn't it a good estimate in this situation? __________
(6)

When you must make absolutely sure you have enough of something, OVERESTIMATE! To overestimate, adjust the numbers by ROUNDING-UP!

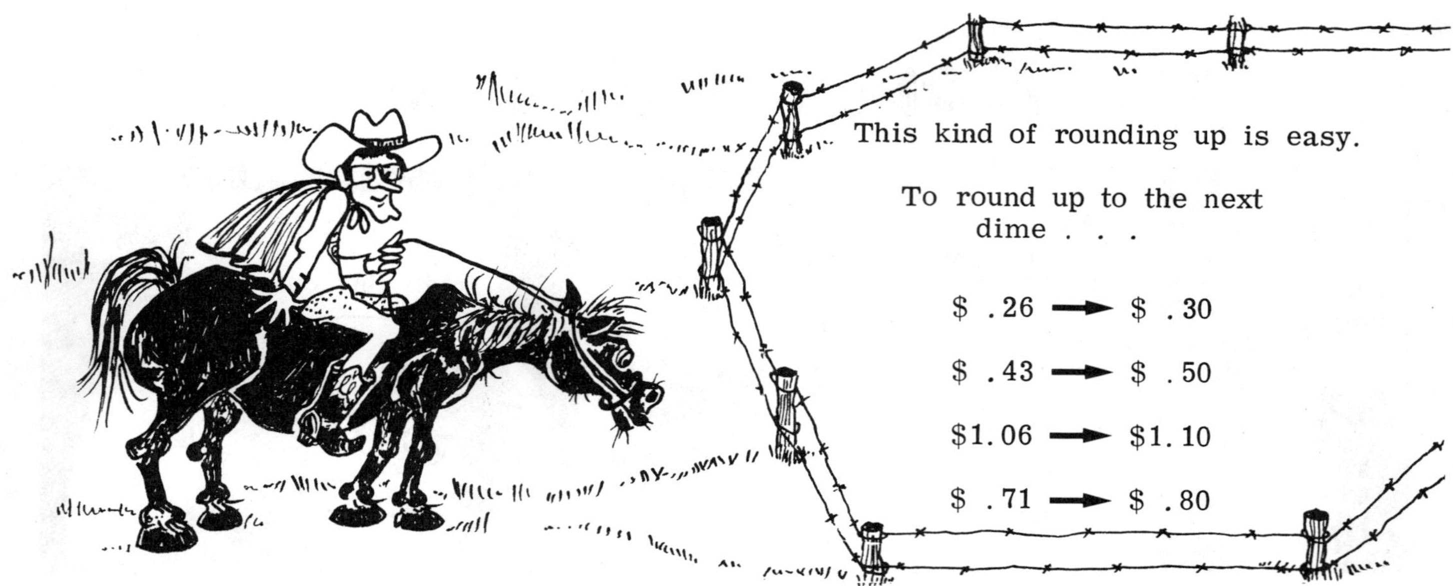

Come and Get It !

To answer the question below, round each amount up to the nearest dime. Find each answer in the Code Box and put the corresponding letter in the blank above the problem number. The first one is done for you.

Who is the most important person at a round-up ?

$$\underline{}\quad \underset{1}{\overset{H}{\underline{}}}\quad \underset{8}{\underline{}}\quad \underset{6}{\underline{}}\quad \underset{2}{\underline{}}\qquad \underset{3}{\underline{}}\quad \underset{10}{\underline{}}\quad \underset{7}{\underline{}}\quad \underset{9}{\underline{}}\quad \underset{5}{\underline{}}$$

1. $ .17 ⟶ $ *.20*

2. $ .63 ⟶ $ _____

3. $1.84 ⟶ $ _____

4. $ .09 ⟶ $ _____

5. $ .55 ⟶ $ _____

6. $2.28 ⟶ $ _____

7. $ .39 ⟶ $ _____

8. $ .43 ⟶ $ _____

9. $1.13 ⟶ $ _____

10. $ .99 ⟶ $ _____

Code Box	C	H	G	U	N	K	A	O	W	C
	.10	.20	.40	.50	.60	.70	1.00	1.20	1.90	2.30

Round up these amounts to the
<u>next half dollar.</u>

$ 2.43 ➡ $2.50

$10.86 ➡ _______ (7)

$ 7.24 ➡ _______ (8)

$13.81 ➡ _______ (9)

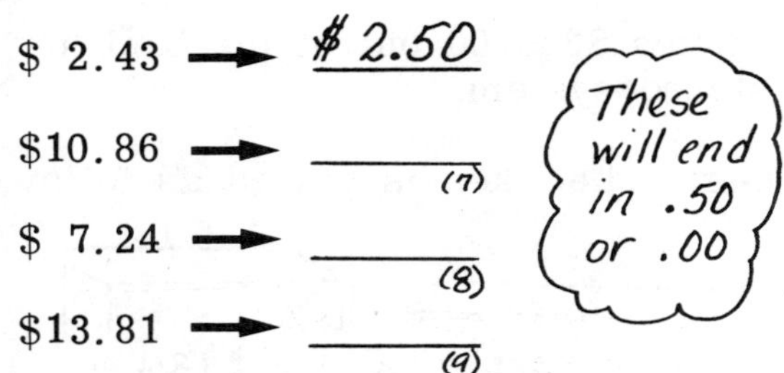

Round up these amounts to the
<u>next dollar.</u>

$74.17 ➡ _______ (10)

$28.06 ➡ _______ (11)

$ 1.93 ➡ _______ (12)

$56.50 ➡ _______ (13)

To find out what this picture is, round up each amount.   Find each answer in the
Code Box and put the corresponding letter in the blank above the problem number.
Some numbers may appear more than once.

<u>4</u> <u>1</u> <u>9</u> <u>7</u> <u>4</u> <u>6</u> <u>2</u> <u>8</u> <u>11</u> <u>3</u> <u>5</u> <u>8</u> <u>10</u> <u>11</u>

1. $ 27.48 ➡ $_______

2. $ 52.84 ➡ $_______

3. $ 71.49 ➡ $_______

4. $117.21 ➡ $_______

5. $ 36.27 ➡ $_______

6. $ 9.53 ➡ $_______

7. $173.81 ➡ $_______

8. $443.28 ➡ $_______

9. $371.24 ➡ $_______

10. $ 63.93 ➡ $_______

11. $590.73 ➡ $_______

Code Box	S	T	H	L	R	G	A	X	E	O	N
	10.00	27.50	36.50	53.00	64.00	71.50	117.50	174.00	372.00	444.00	591.00

 The Money Round-Up 3

Round up and OVERESTIMATE
to get the answers to the
following problems.

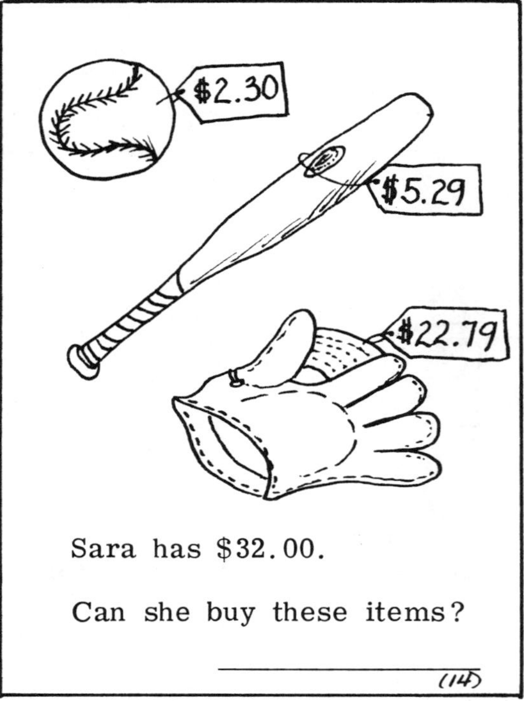

Sara has $32.00.

Can she buy these items?

_________________ (14)

Juanita takes $350.00 out of the bank to
buy a stereo system.

The system she likes is priced as follows:

Does she have enough money to buy
this system?

_________________ (15)

Jacob's mom gives him a ten
dollar bill and tells him,
"Go to the store and buy milk,
hamburger, carrots, potatoes
and sugar. If there is enough
left over, buy yourself a treat."

Here are the prices of the food Jacob bought:

Milk	$ .99
Hamburger	$1.89
Carrots	$ .47
Potatoes	$2.43
Sugar	$3.59

Will there be enough money left over for
Jacob to buy a $ .25 pack of gum? _________ (16)

a $ .75 comic book? _________ (17)

          The Money Round-Up   4

There are other situations besides money where you should OVERESTIMATE by rounding-up. These are situations where you want to make sure you have enough of something.

Estimating the amount of paint needed to paint a room.

Estimating the wallpaper needed to cover a wall.

Estimating the material needed to make clothing.

Estimating the carpeting needed to cover a floor.

Estimating the quantity of lumber needed for a project.

Estimating the amount of food necessary to entertain guests at a party.

Amelia is going to make matching vests for each person in her band.

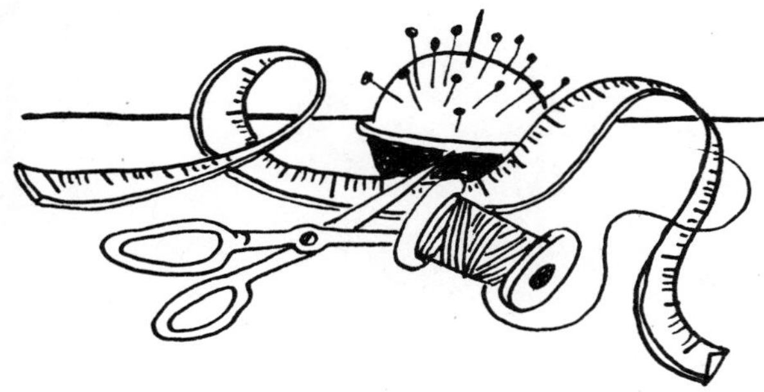

There are 6 people in the band. Each vest takes 3/8 **yard**.

There a 3 pieces of fabric to choose from. Which should she buy to be sure to have enough fabric.

    a)   2 yards

    b) $2\frac{1}{2}$ yards

    c)   3 yards

(18)

Wendy has invited 27 friends to a potluck supper. She will provide plates and pop.

She will serve food on paper plates which come 10 to a package. How many packages does she need to buy to have enough plates? ________

(19)

The party book says to allow between 12 and 15 ounces of pop for each person at the party.

There are 32 ounces in a quart. How many quarts of pop should she buy to have enough? ________

(20)

1. $1.00
2. yes
3. $1.10
4. no
5. $1.02
6. she didn't have enough money for lunch
7. $11.00
8. $7.50
9. $14.00
10. $75.00
11. $29.00
12. $2.00
13. $57.00
14. yes
15. no
16. yes
17. no
18. c) 3 yards
19. 3
20. 14 or 15

PUZZLE ANSWERS

page 2   (Message Puzzle):
CHUCK WAGON

page 3   (Message Puzzle):
A TEXAS LONGHORN

general math project

minneapolis public schools

UNIT: GREAT ESTIMATIONS

DESCRIPTION: This unit consists of 66 measurement estimation activity cards, two cards to a page. They deal with a variety of topics that are coded as follows:

ESTIMATION TOPIC	CARD NUMBERS
How Many	1A-14A
Length	1B-16B
Height	1C-10C
Weight	1D-8D
Temperature	1E-7E
Area	1F-6F
Time	1G-5G

Each section has a Guide Sheet to assist you or your students with the problems and to provide answers.

These activity cards are designed to be used as short fillers for the end of the class or for short periods of time when you want to reinforce estimation skills.

You may wish to print these pages on heavy stock and laminate them (after you have cut them apart) to be used as a classroom set.

SKILLS TO BE REINFORCED: measurement estimation

In the <u>HOW MANY</u> section of these estimation cards, there is a general strategy that is used. You need to take a sample, count or estimate the number of objects in that sample, estimate how many of the sample size are in the whole, and then multiply to estimate the total number. Part of the learning experience in this section may be to have students devise their own strategies in order to arrive at a good estimate.

CARD
NUMBER

1. Answers will vary. Use the <u>HOW MANY</u> strategy.

   Count the number of beans in the lowest layer, estimate the number of layers and then multiply.

2. The actual number of hairs is 800. Use the <u>HOW MANY</u> strategy.

3. Answers will vary. Use the <u>HOW MANY</u> strategy.

4. There are actually 16 fishes and 6 birds.

5. In the '82 Mpls. Directory there are about 100 names per column and about 92 columns. Answers will vary in other locales.

6. There are about 70 bumps. Use the <u>HOW MANY</u> strategy.

7. The answers will vary according to hand size. About 25 or more.

8. This should be a classroom activity. Use the <u>HOW MANY</u> strategy.

9. There are about 5 gallons in a round wastebasket that has a 12 inch diameter and is 14 inches high.

10. There are between 10 and 11.

    A small juice can is 6 fl. oz. and a half gallon milk carton is 64 fl. oz.

11. This is a trick question. There is no dirt in a hole.

12. Answers will vary. Use the <u>HOW MANY</u> strategy.

13. Answers will vary. Use the <u>HOW MANY</u> strategy.

14. If you continue to guess halfway between the existing upper and lower limits, you should be able to guess the number in less than 8 guesses.

Fill a pint jar with beans.

ESTIMATE HOW MANY BEANS
ARE IN THE JAR.

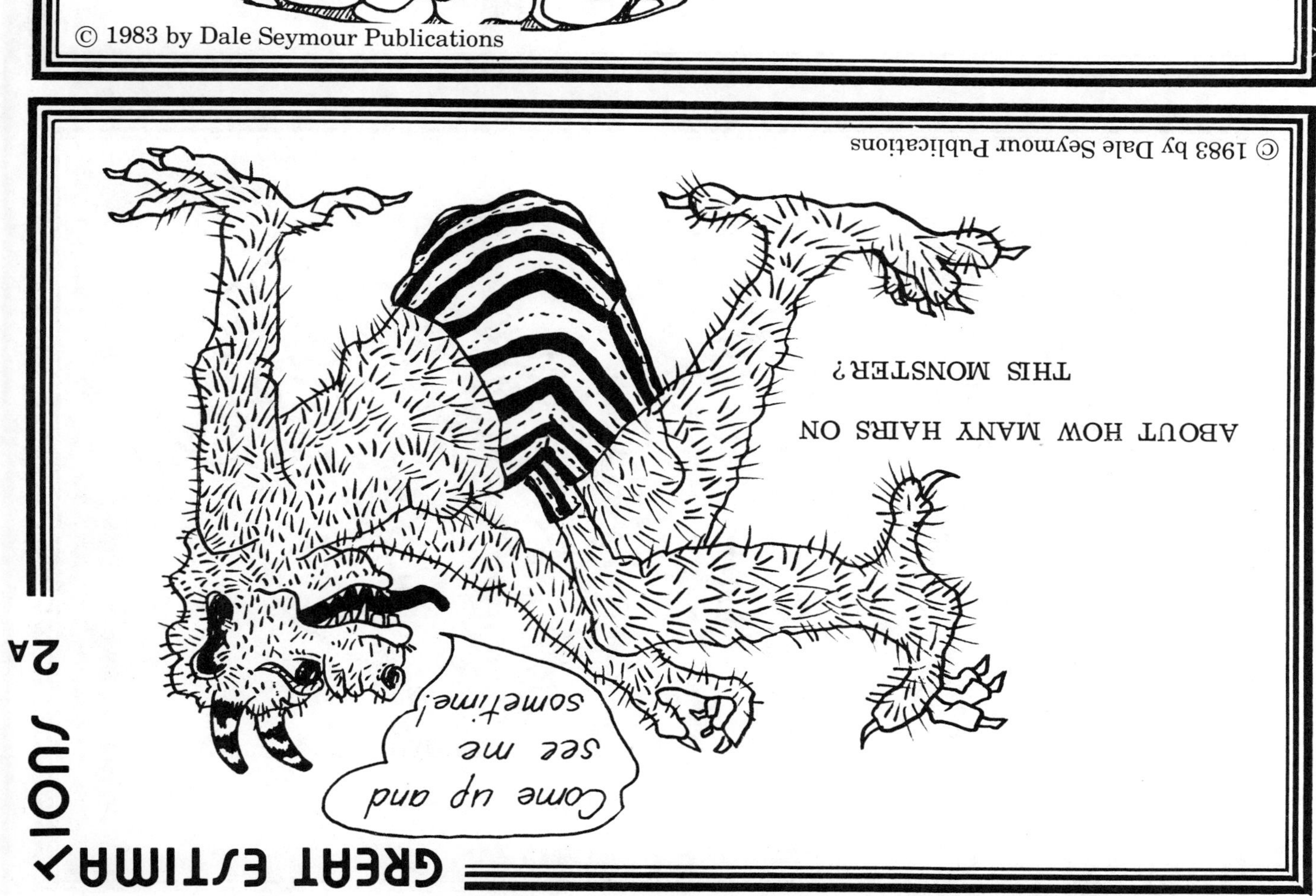

ABOUT HOW MANY HAIRS ON
THIS MONSTER?

GREAT ESTIMATIONS

3A

ESTIMATE THE NUMBER OF WINDOWS IN YOUR SCHOOL BUILDING.

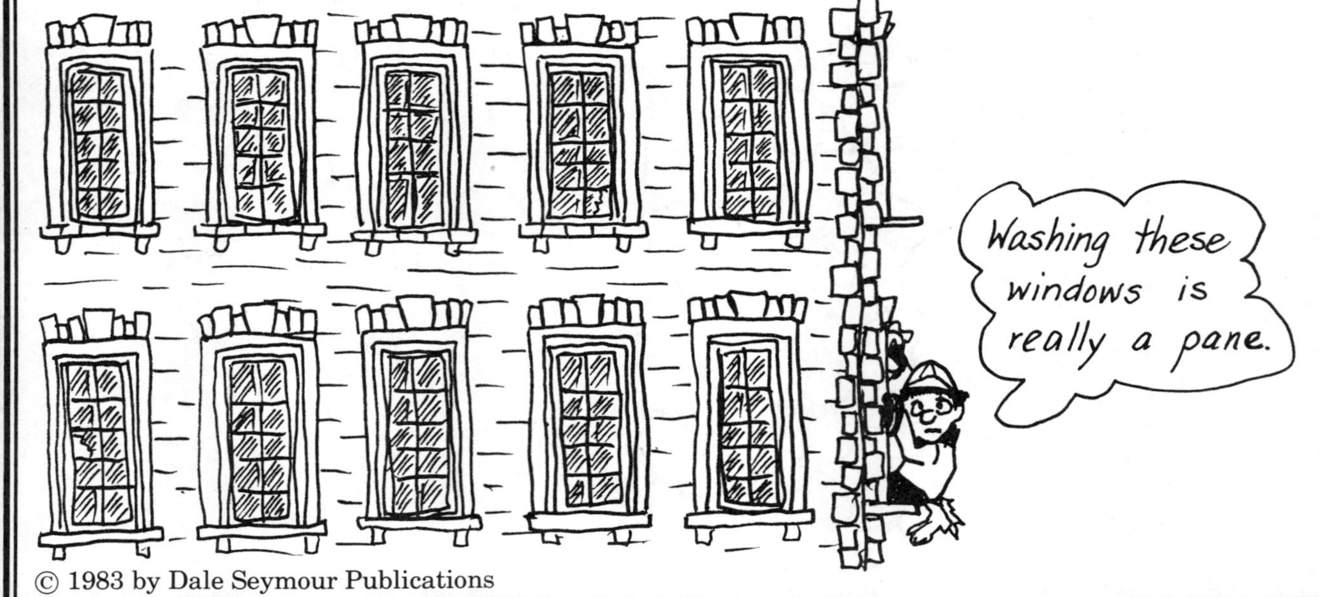

GREAT ESTIMATIONS

4A

ESTIMATE THE NUMBER OF BIRDS YOU SEE.

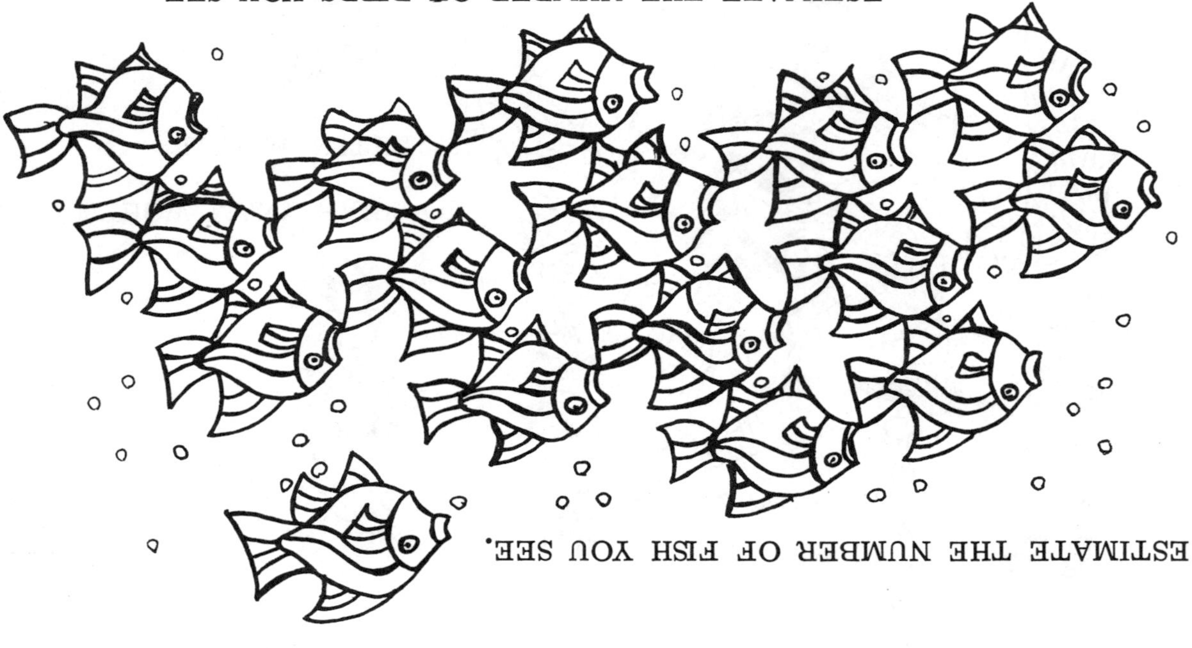

ESTIMATE THE NUMBER OF FISH YOU SEE.

© 1983 by Dale Seymour Publications

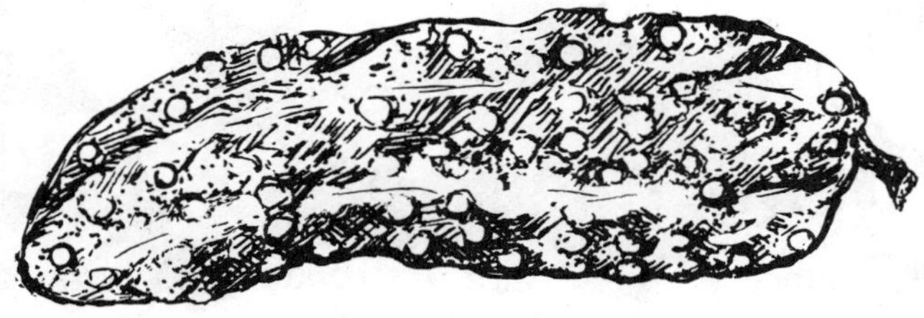

About how many Johnson's are listed in your local telephone book?

Hint: Count the number of listings in a column and multiply!

ABOUT HOW MANY BUMPS ARE ON THIS DILL PICKLE?

GREAT ESTIMATIONS

7A

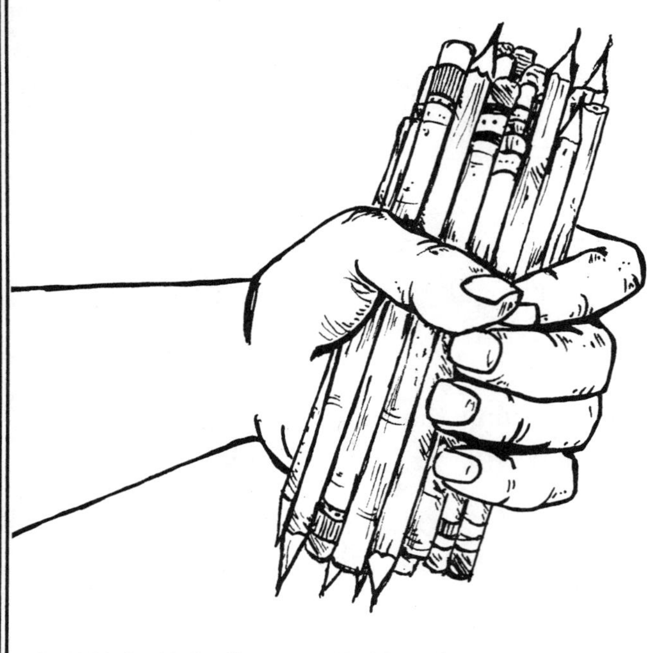

ABOUT HOW MANY PENCILS OR PENS CAN YOU GRASP IN ONE HAND?

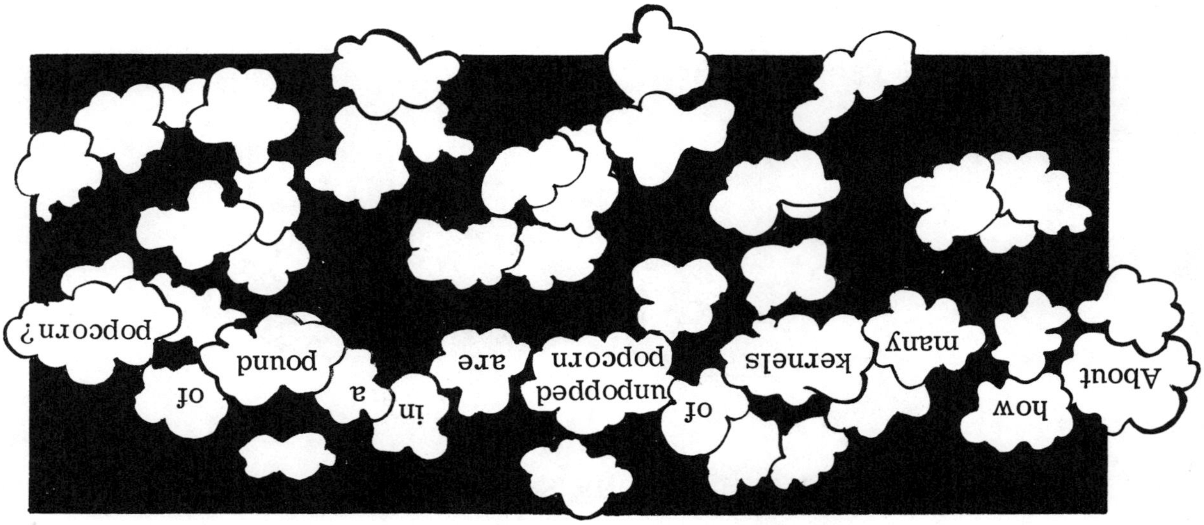
About
how
many
kernels
of
unpopped
popcorn
are
in
a
pound
of
popcorn?

GREAT ESTIMATIONS

8A

ABOUT HOW MANY GALLONS OF MILK WOULD THE WASTEBASKET IN YOUR CLASSROOM HOLD?

HINT: A large milk carton or a plastic jug holds 1 gallon.

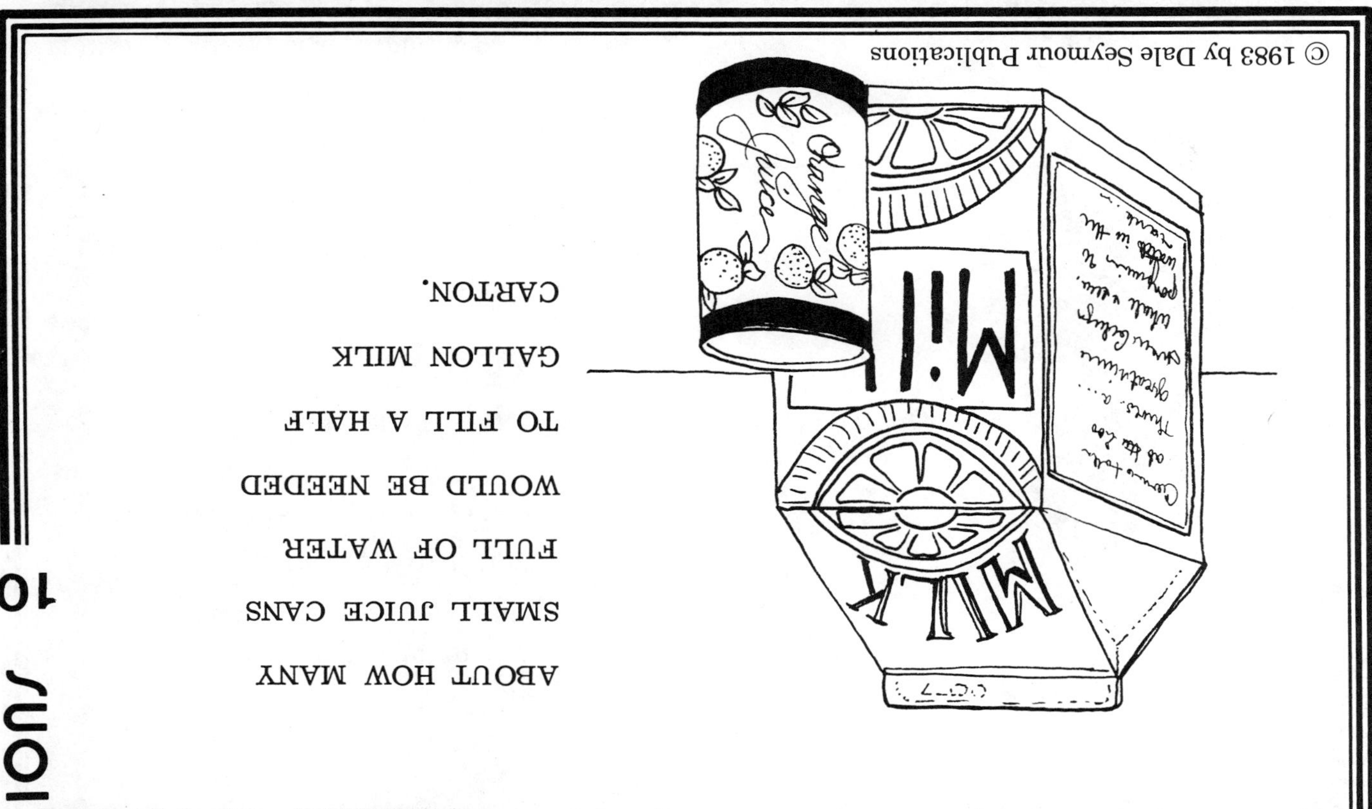

ABOUT HOW MANY SMALL JUICE CANS FULL OF WATER WOULD BE NEEDED TO FILL A HALF GALLON MILK CARTON.

GREAT ESTIMATIONS

11A

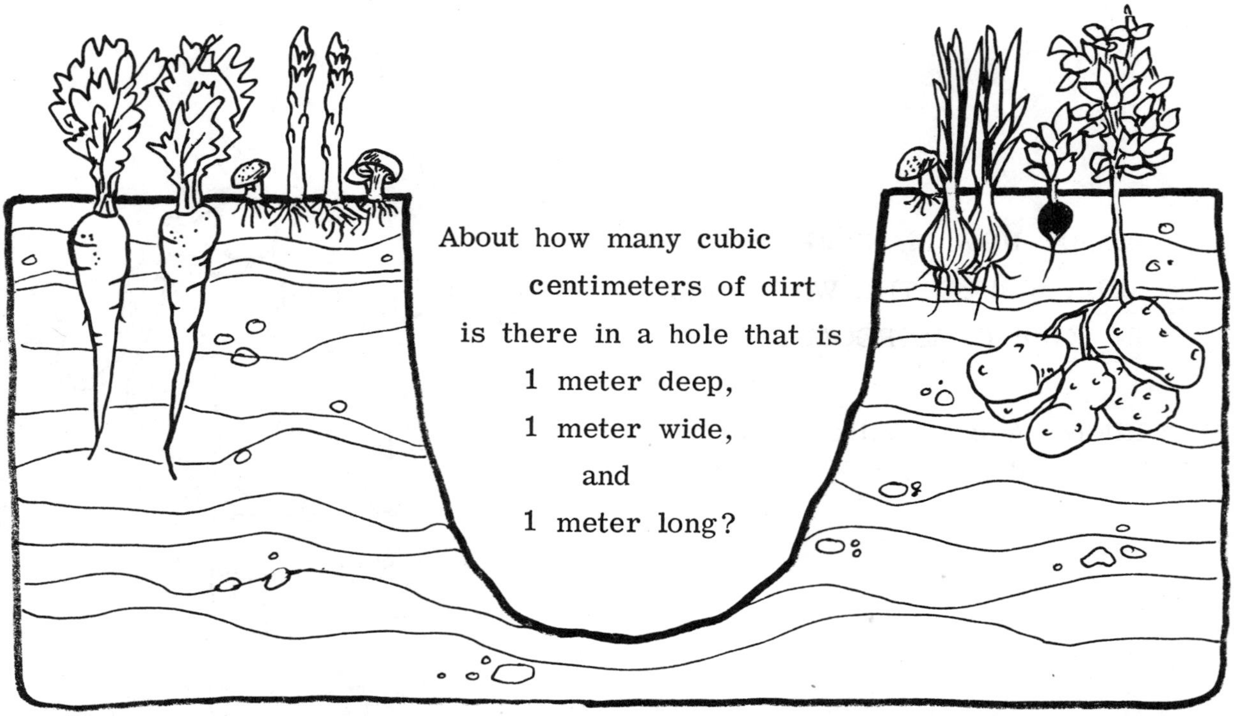

About how many cubic centimeters of dirt is there in a hole that is
1 meter deep,
1 meter wide,
and
1 meter long?

GREAT ESTIMATIONS

12A

USE A PAPERBACK BOOK.

Count the words in a line.
Count the lines on the page.
Estimate the words on a page.
Do this for two more pages.

ESTIMATE HOW MANY WORDS ARE IN THE BOOK.

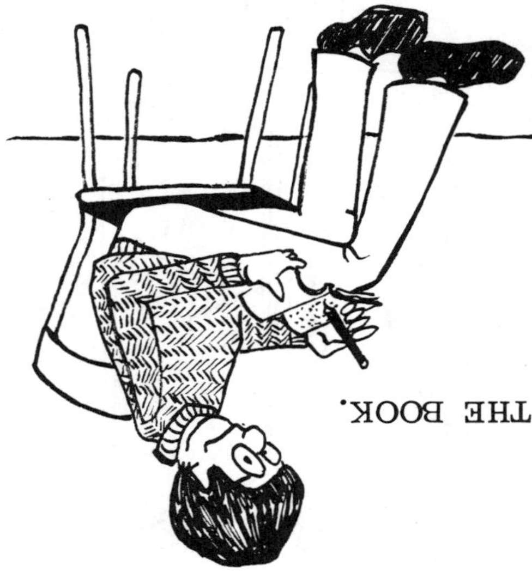

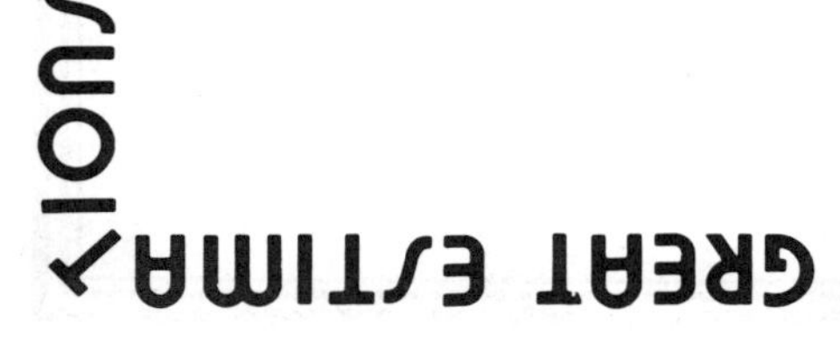

How many steps are there in the stairway from 1ˢᵗ floor to 2ⁿᵈ floor in your school?

How many steps are there in all the stairways in the whole school?

GUESS THE NUMBER GAME
(for two people)

1. You pick a number from 0 to 99.
   Your partner will try to guess the number.

2. When your partner makes a guess, tell him/her "higher" or "lower".

3. Count how many tries it takes to guess the number.

Example: the number is 18

partner's guess: 50   your response: lower
partner's guess: 40   your response: lower
partner's guess: 10   your response: higher
etc. etc. etc.   . . . . .

4. Can you discover another method whereby you can always guess the number in less than 8 guesses?

Length
Cards 1B-16B

CARD NUMBER

1.   The answers spell R I G H T.

2.   In most parts of Minneapolis, north-south blocks are long; 8 to a mile: east-west blocks are short; 16 to a mile. (Answers may vary.)

3.   A sidewalk in a long city block might be 200 yards long. For example, if the squares are a yard or so square, 150-250 might be a reasonable answer.

4.   About 6 pencils.

A new pencil is about $7\frac{1}{2}$ inches or 19 cm. A desk is about 4 feet long or 120 cm.

5.   Answers will vary.

6.   Answers will vary.   A teacher/classroom activity.

7.   Answers will vary.   A teacher/classroom activity.

8.   Answers will vary.

A standard basket is about 3 decimeters across.

9.   Answers will vary.

The average hand span is about 7-9 inches.

10.   Answers will vary.

It might be helpful to have string available to cut to the length of the reach in order to compare it to the height. You could also use a ruler to measure marks on the chalkboard.

11.   About 27 or 28 cm by 21 or 22 cm.

12.   Answers will vary.

13.   Answers will vary.

14.   Answers will vary.

15.   A teacher/classroom activity.   Display several cylindrical objects, most of which will have a greater circumference.

16.   Computer activity.

Great Estimations

0   ↑ G   ↑ H   ↑ R   ↑ I   ↑ T   1

1. Which arrow is about $\frac{1}{2}$ the distance from 0 to 1?   ???

2. Which arrow is about $\frac{2}{3}$ the distance from 0 to 1?   ???

3. Which arrow is about $\frac{1}{5}$ the distance from 0 to 1?   ???

4. Which arrow is about $\frac{3}{8}$ the distance from 0 to 1?   ???

5. Which arrow is about $\frac{3}{4}$ the distance from 0 to 1?   ???

ABOUT HOW MANY LONG CITY BLOCKS ARE THERE IN A MILE?

ABOUT HOW MANY SHORT CITY BLOCKS ARE THERE IN A MILE?

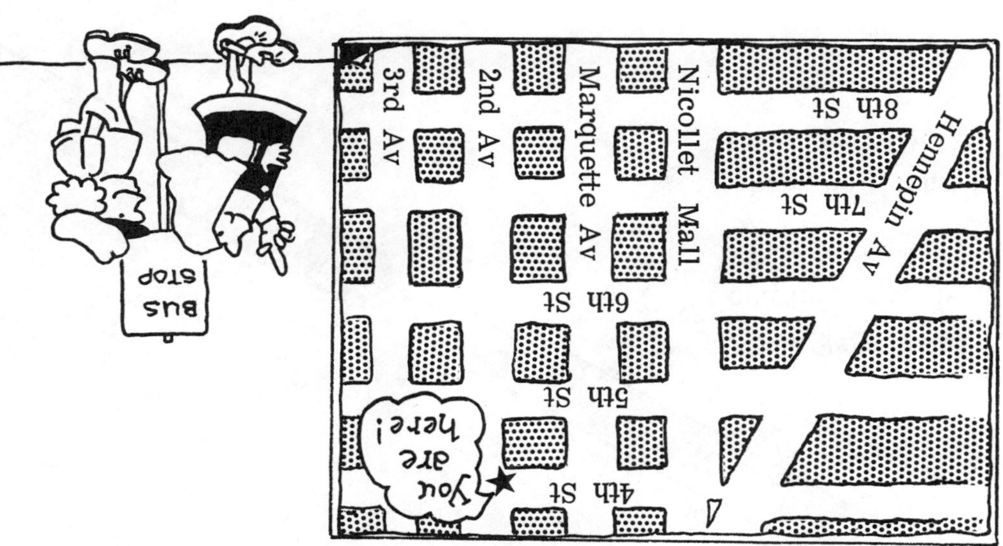

ABOUT HOW MANY SQUARES
ARE THERE IN THE SIDEWALK
IN A CITY BLOCK?

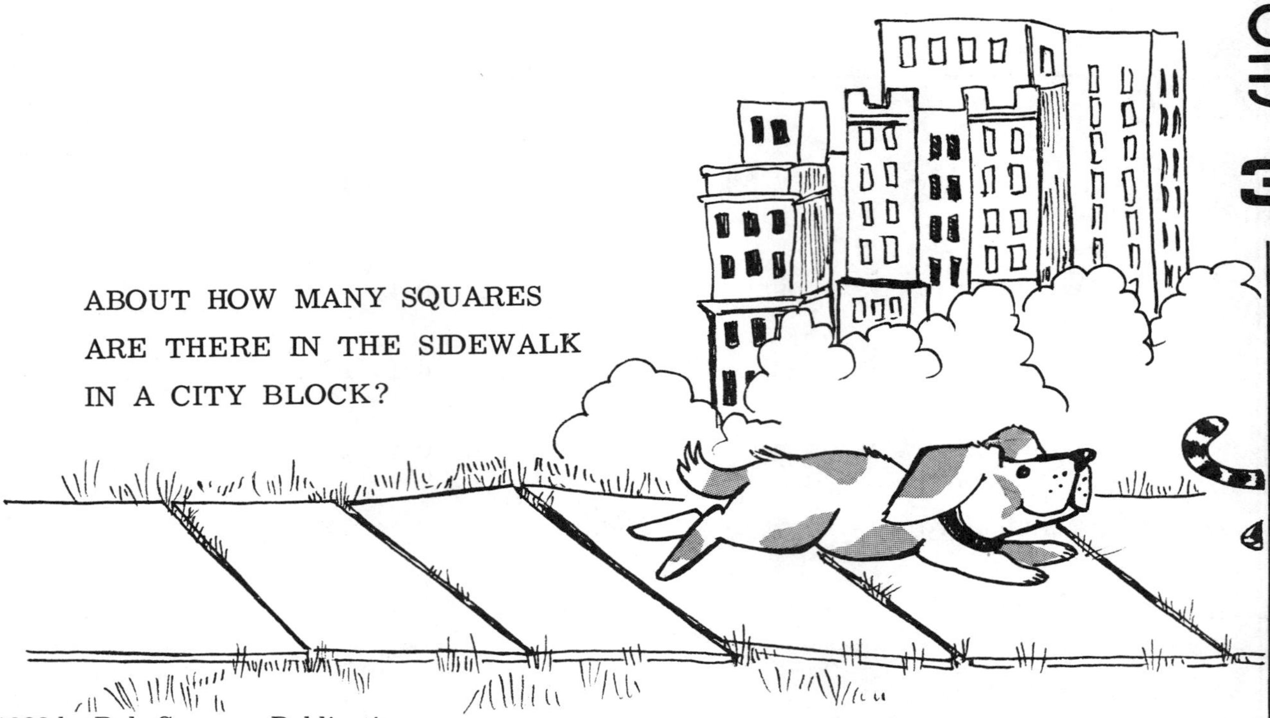

About how long, in inches or
centimeters, is a new pencil?

About how long is your
teacher's desk?

How many new pencils laid end to end
would equal the length of your teacher's
desk?

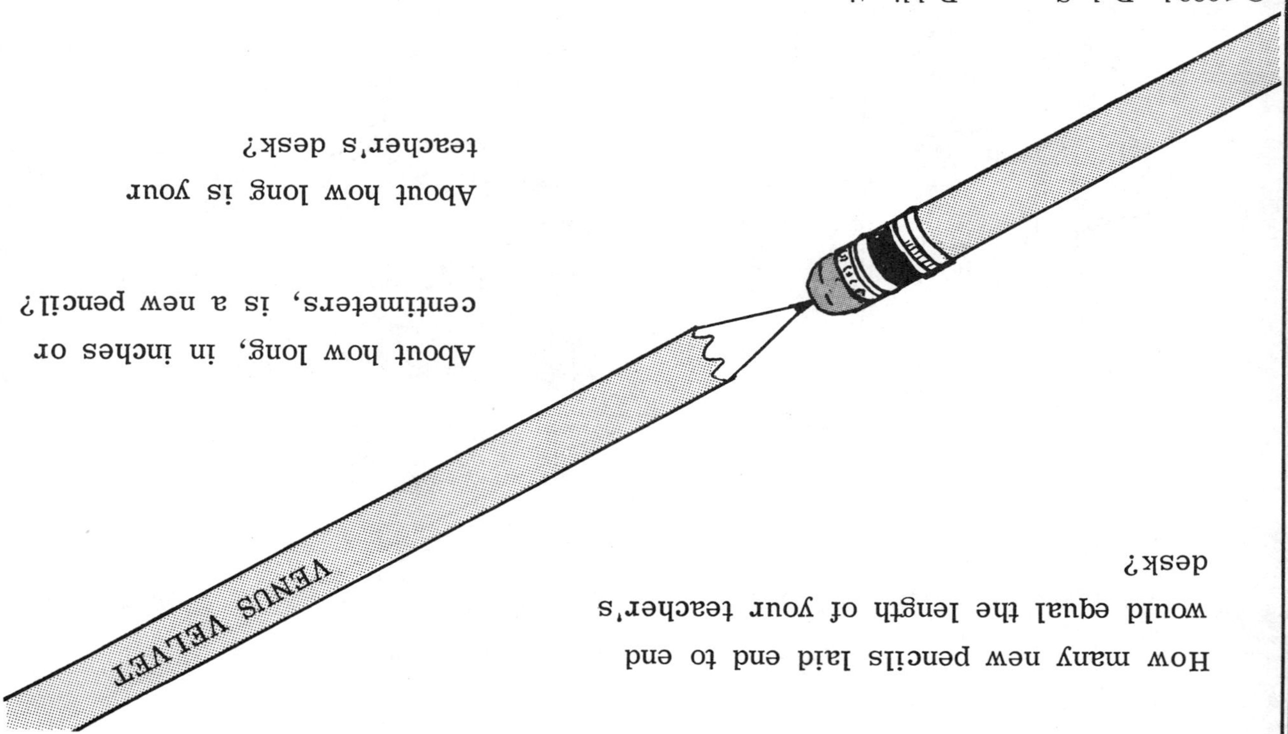

GREAT ESTIMATIONS 5B

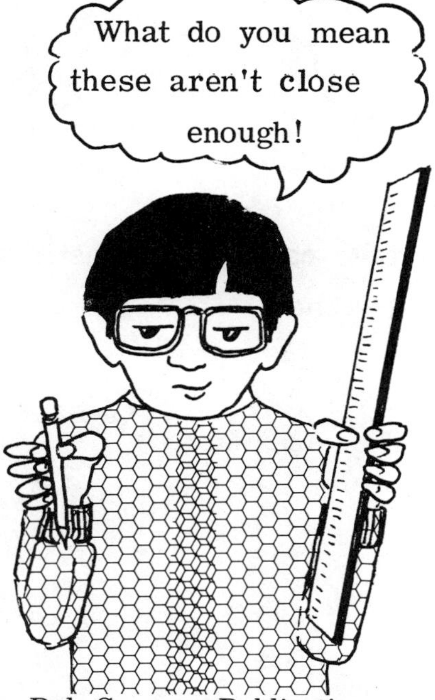

FIND TWO DIFFERENT THINGS IN THE CLASSROOM THAT ARE ABOUT THE SAME LENGTH.

GREAT ESTIMATIONS 6B

DRAW A LINE ON THE BLACKBOARD.

Have each student estimate its length.

Then measure the line.

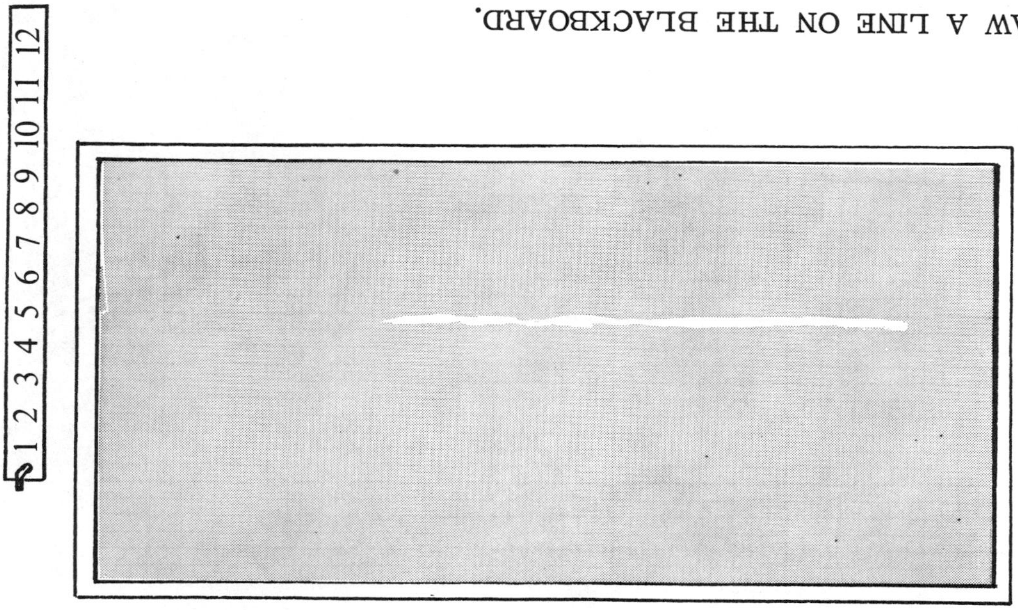

LAY A PIECE OF STRING ON A TABLE,
SOMETHING LIKE THIS:

Have each student estimate
its length.

Then measure.

THE WASTEBASKET IN
YOUR ROOM IS ABOUT
HOW MANY DECIMETERS
ACROSS?

This is a decimeter.

MEASURE YOUR HAND SPAN (the distance from thumb tip to little finger tip, spread out).

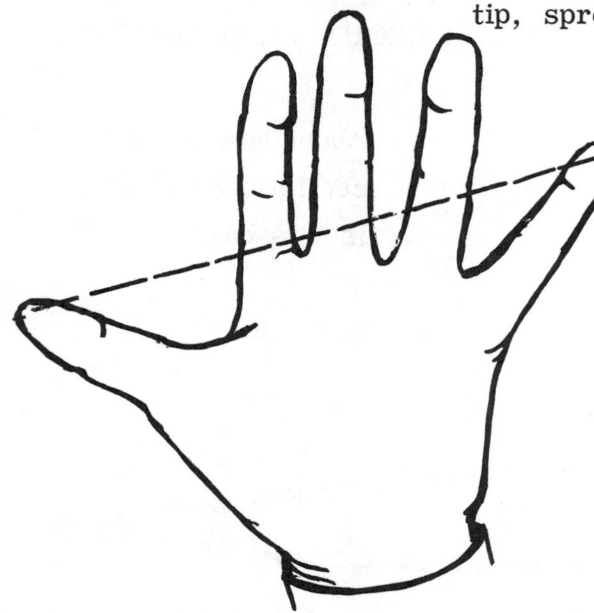

How many spans is the length of your desk?

Estimate, in inches, the length of your desk.

How many spans is the height of the chalkboard?

Estimate, in inches, the height of the chalkboard.

If you stretch out your arms, is your reach (distance from fingertip to fingertip) longer or shorter than your height?

Would you fit into a square or a rectangle?

11B

THE WIDTH OF YOUR INDEX FINGERNAIL IS ABOUT 1 CENTIMETER.

About how many centimeters long is this card?

How wide?

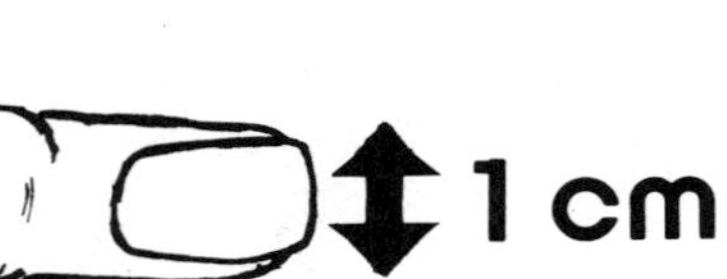

12B

MEASURE YOUR PACE (IN FEET) FROM TOE TO TOE.

Walk from one end of the room to the other.

About how many paces is the length of the room?

About how many feet is the length of the room?

About how many paces is it from the classroom door to the nearest drinking fountain?

About how many feet would that be?

Pace the length of the hallway.

Use this to estimate the length of the school building.

13B

Guess the
length of
your hair.

CHECK:

Pull out one
of your hairs
and measure
it.

© 1983 by Dale Seymour Publications

14B

© 1983 by Dale Seymour Publications

Cut a piece of string to a
length that you think is the
same as the circumference
of your head.

Does it fit around your
head?

GREAT ESTIMATIONS 15B

FIND A ROUND CAN, GLASS, JAR, ETC., WHOSE HEIGHT IS GREATER THAN ITS CIRCUMFERENCE.

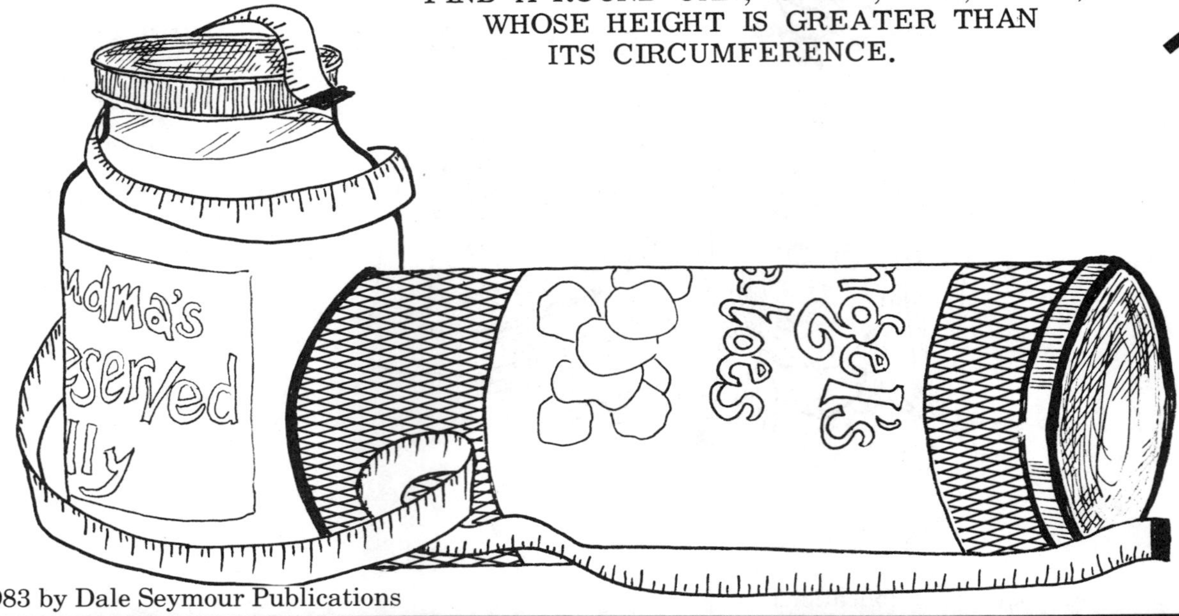

GREAT ESTIMATIONS 16B

If your school has an APPLE computer, play the estimation game "Darts" on it.

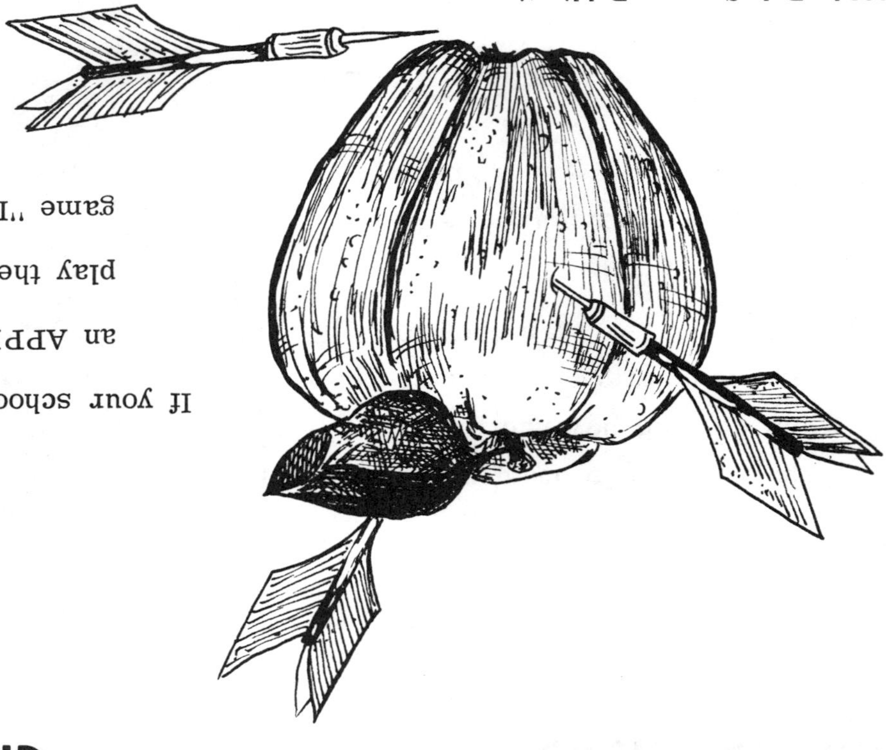

CARD
NUMBER

1.   Results will vary.

2.   Results will vary.

3.   Results will vary.

4.   About 3 times.

5.   The height of five stacked nickels does not equal the diameter of one quarter, so the nickels would be a better deal.

6.   Check with the school engineer. He may be able to tell you the actual height.

7.   Answers will vary.

8.   About 13 stacked pennies equal a penny's height on edge.

9.   About 3 3/4 - 4 inches.

10. About 8 inches.
     A 2 inch ream of paper contains 500 sheets of paper.

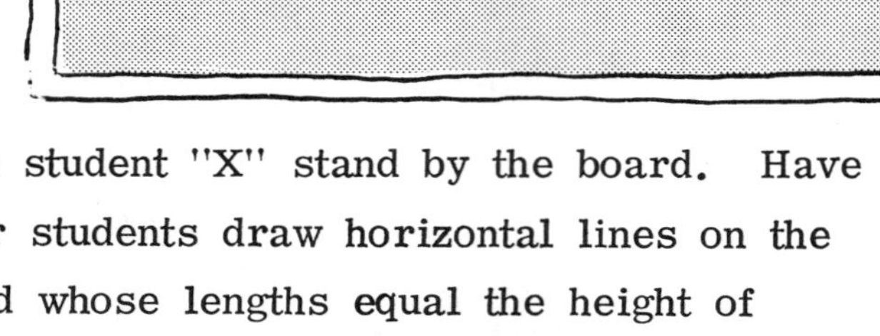

Have student "X" stand by the board. Have other students draw horizontal lines on the board whose lengths equal the height of student "X". Then measure to check.

Have student "X" stand up. Have other students put marks on the board to estimate the height of student "X". Then have student "X" stand by each mark to see whose estimate is closest.

GREAT ESTIMATIONS 3c

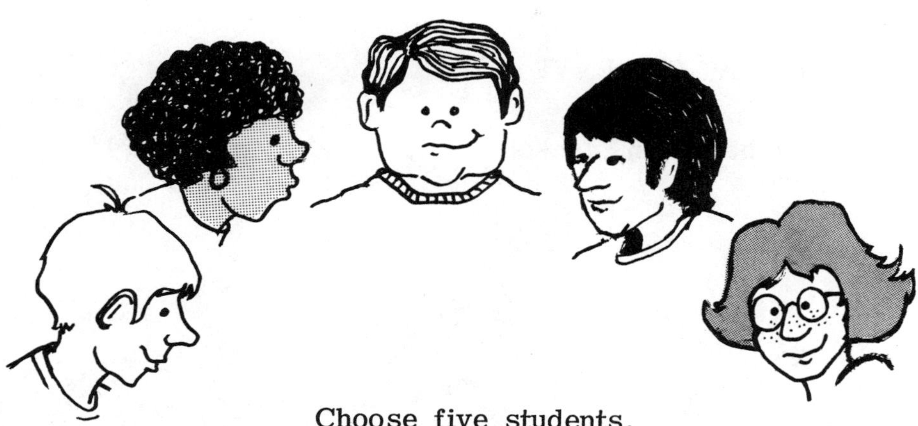

Choose five students, who are to remain seated. Have the class list their five names in order of height. Then have the five students line up in order of height.

GREAT ESTIMATIONS 4c

If you had a string as long as your height, about how many times could you wrap it around your head?

TRY IT.

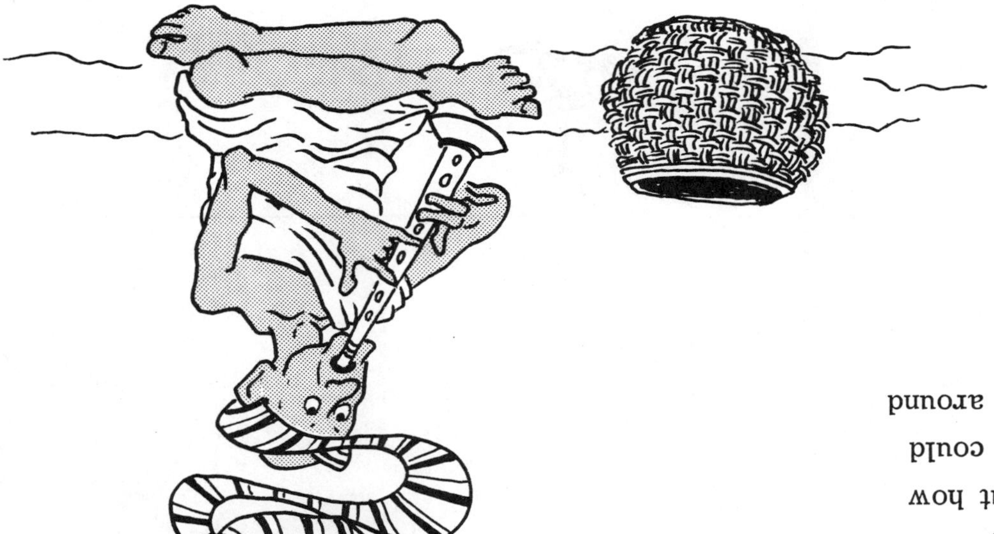

GREAT ESTIMATIONS 5c

WOULD YOU RATHER HAVE . . .

a) your height in stacked nickels

OR

b) your height in quarters placed end to end

GREAT ESTIMATIONS 6c

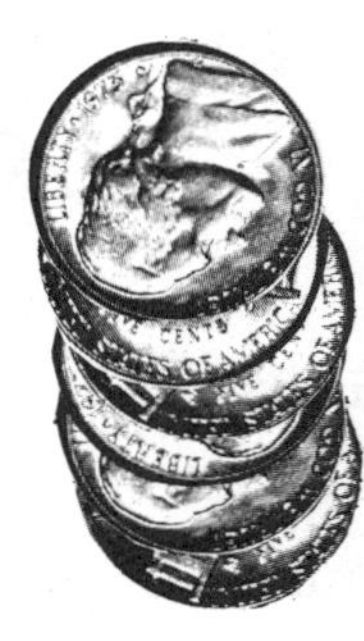

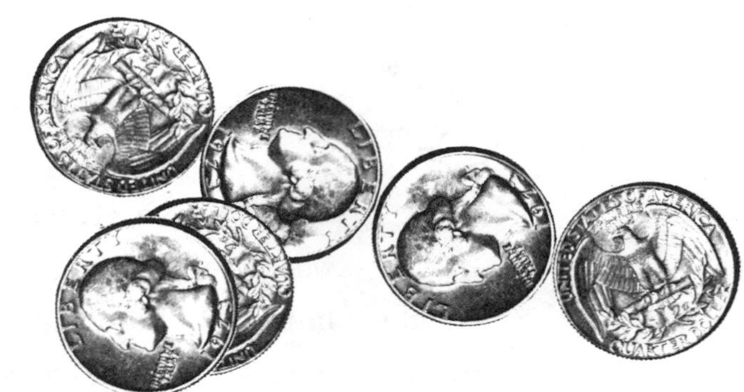

IF YOUR SCHOOL HAS A FLAGPOLE, ESTIMATE ITS HEIGHT.

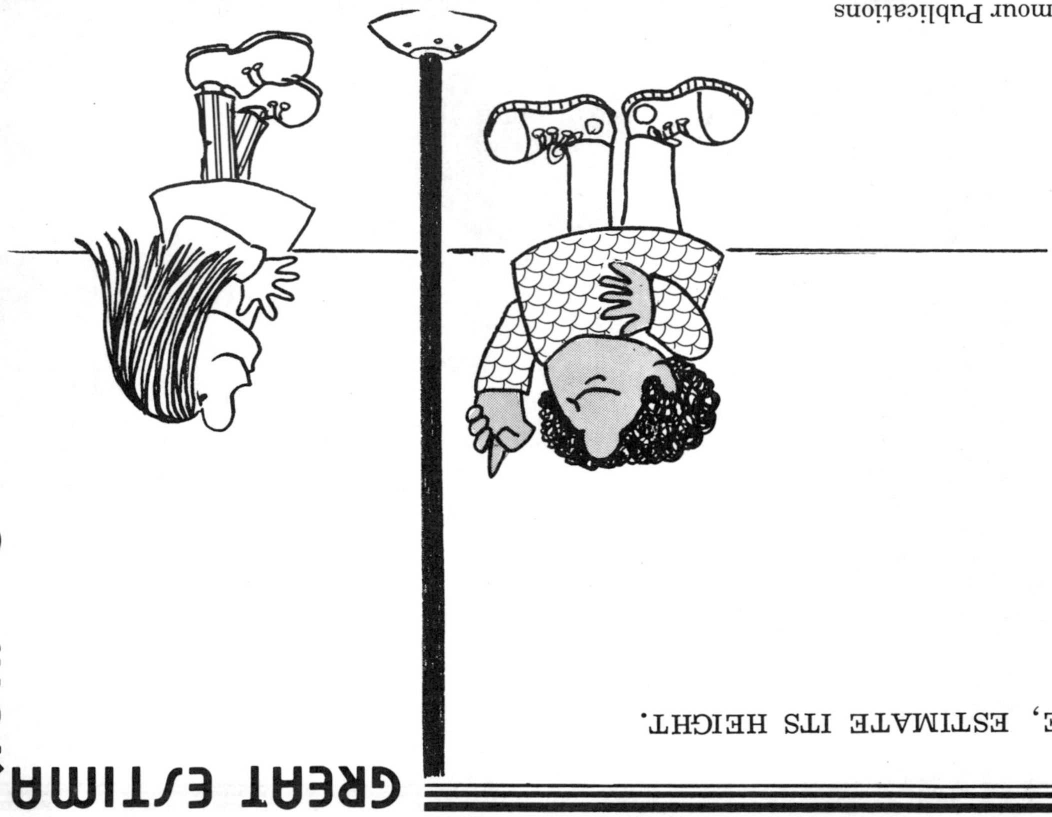

Estimate the height of the classroom ceiling.

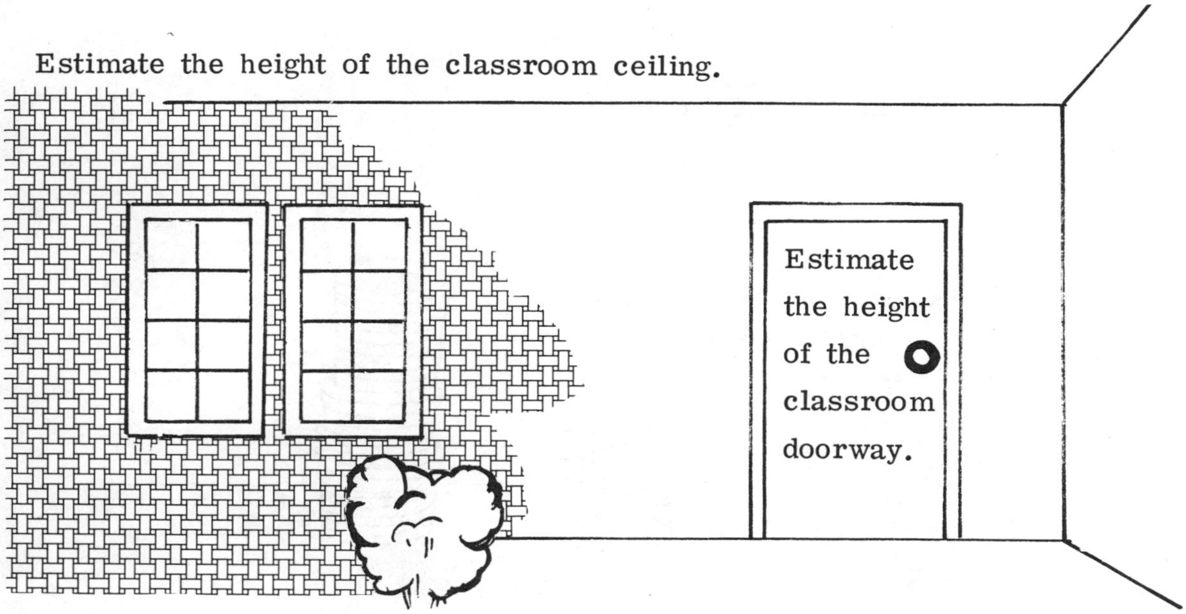

Estimate the height of the classroom doorway.

Estimate the height of the school.

© 1983 by Dale Seymour Publications

© 1983 by Dale Seymour Publications

ESTIMATE HOW MANY PENNIES IT WOULD TAKE TO MAKE A STACK AS HIGH AS ONE PENNY STANDING ON EDGE.

TRY IT.

ABOUT HOW HIGH IS A STACK OF
50 PENNIES?

About how high is a stack of 2000 sheets of paper?

CARD
NUMBER

1. The answer would be about 12-17 people.

   The average ninth grader weighs about 120 lbs.
   The average adult weighs about 150 lbs.

2. Answers will vary.

   Be sure to not embarass a student who does not want to talk about his/her weight.

3. Answers will vary.

   It would be helpful to have a scale in the classroom or at least some individual items that weigh 5 lbs. or 10 lbs.

4. Answers will vary.

5. A textbook weighs about 1.5 - 2 lbs.

6. A new piece of chalk weighs about 15-20 g.

7. Answers will vary.

8. A Minneapolis telephone book weighs about 3.5 - 4 lbs.

A sign in an elevator says:

ABOUT HOW MANY PEOPLE CAN RIDE SAFELY?

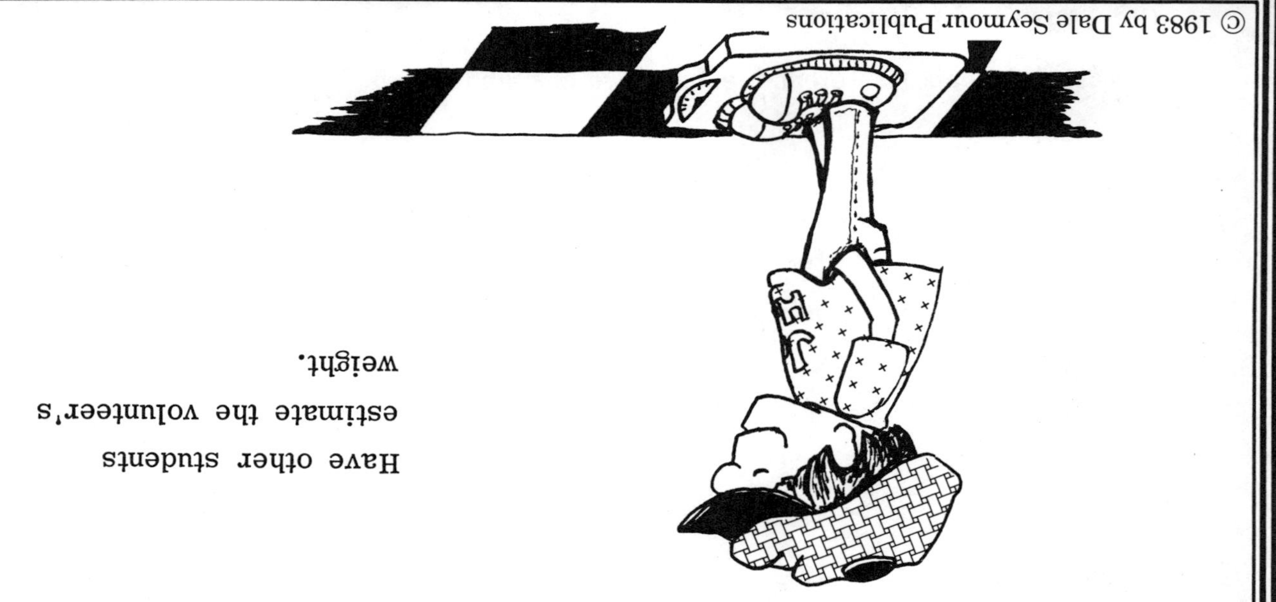

Have other students estimate the volunteer's weight.

Ask for a student who will volunteer to tell you his/her weight.

Collect a number of items in the classroom that you estimate to weigh a total of 5 pounds.

Try 10 pounds.

Check by using a scale.

© 1983 by Dale Seymour Publications

© 1983 by Dale Seymour Publications

FIND TWO DIFFERENT THINGS IN THE CLASSROOM THAT WEIGH ABOUT THE SAME.

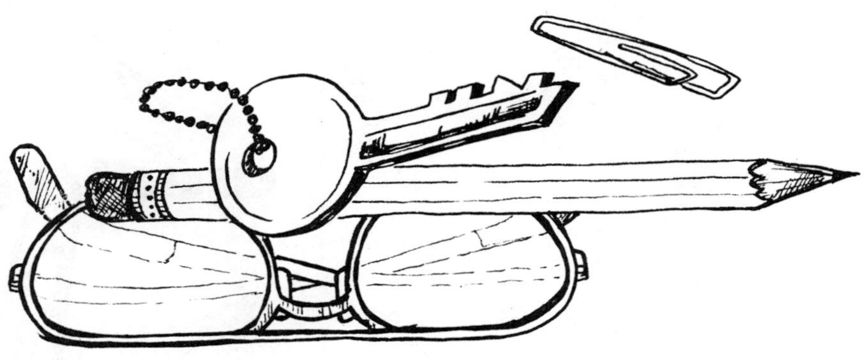

ESTIMATE THE WEIGHT OF A TEXTBOOK.

How many textbooks would weigh about 10 pounds?

about 15 pounds?

about 20 pounds?

If the weight of two paper clips is 1 gram, about how many grams does a new piece of chalk weigh?

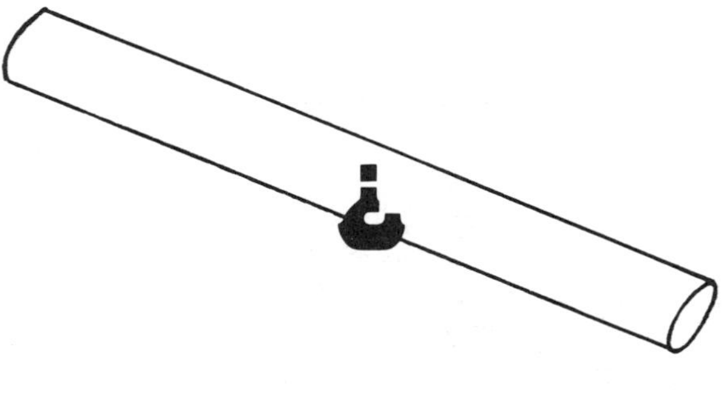

GREAT ESTIMATIONS 7D

Choose five items in your classroom and list them in order from lightest to heaviest.

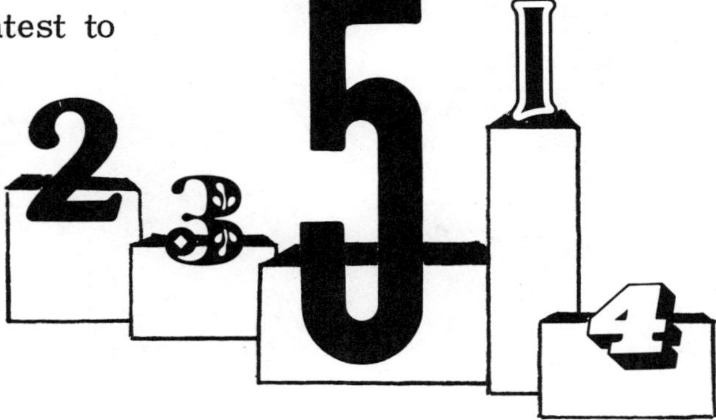

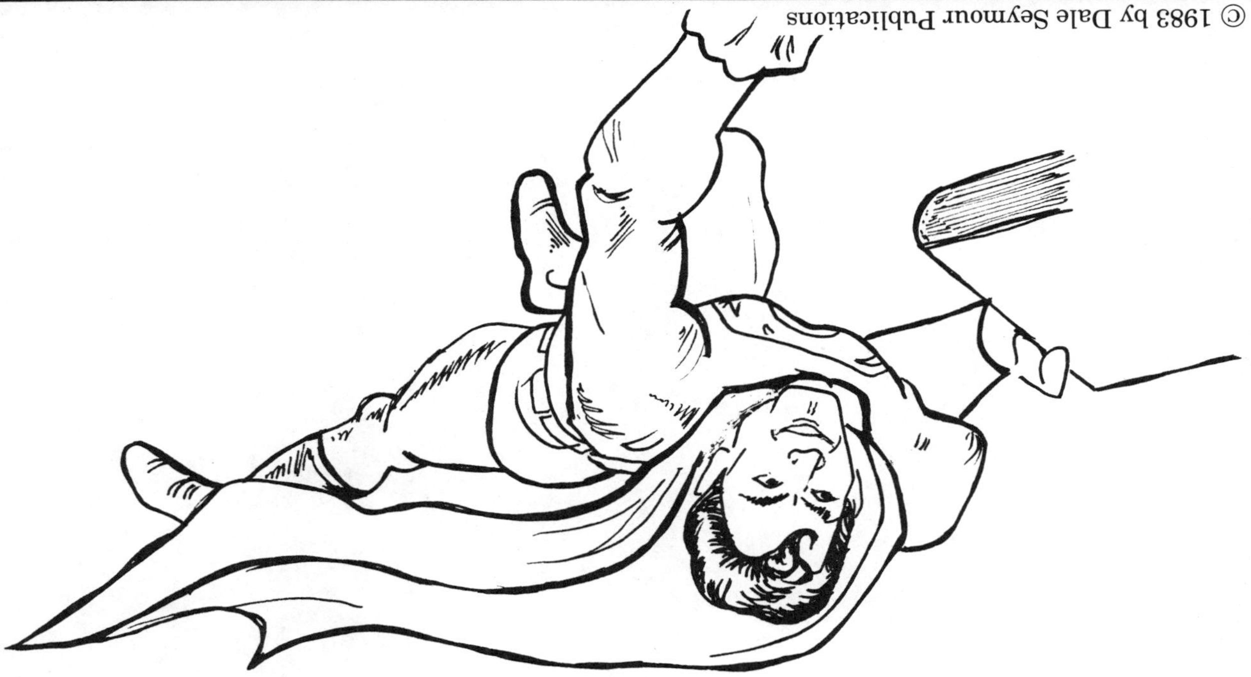

ESTIMATE THE WEIGHT OF A TELEPHONE BOOK.

GREAT ESTIMATIONS 8D

CARD
NUMBER

1.

	F	C
Cold	below 32°	below 0°
Cool	32° – 60°	0° – 15°
Warm	60° – 75°	15° – 22°
Hot	above 75°	above 22°

2. Below freezing

3. Answers will vary.

4. Here is a guide for you.

C	-40	-30	-20	-10	0	10	20	30
F	-40	-22	-4	14	32	50	68	86

5. See #4 above for help in conversion from C to F.

6. $30°$ C $\approx 86°$ F. It would be hot.

7. About $-20°$ C to $5°$ C or $-5°$ F to $38°$ F.

Give a temperature that you think is...

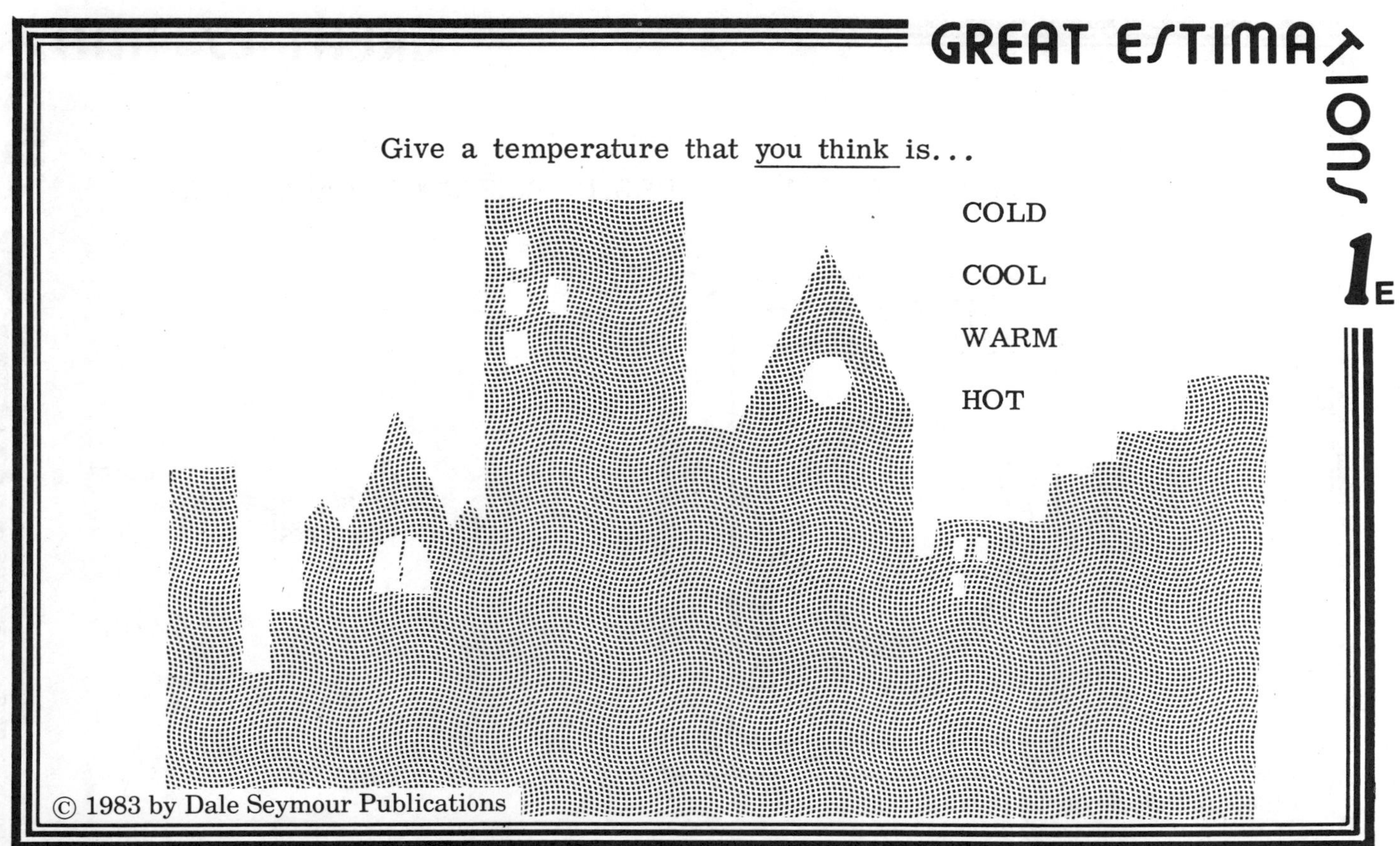

COLD

COOL

WARM

HOT

ESTIMATE
THE TEMPERATURE
OF AN
ICE CUBE.

GREAT ESTIMATIONS 3E

ESTIMATE THE TEMPERATURE IN THIS ROOM RIGHT NOW.

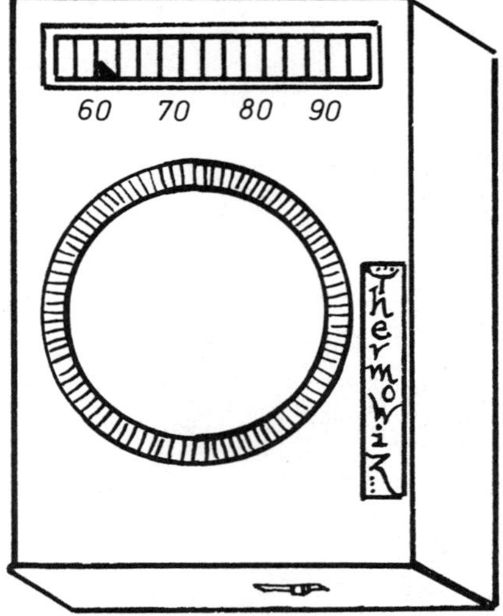

Check the thermostat to see how close you are.

GREAT ESTIMATIONS 4E

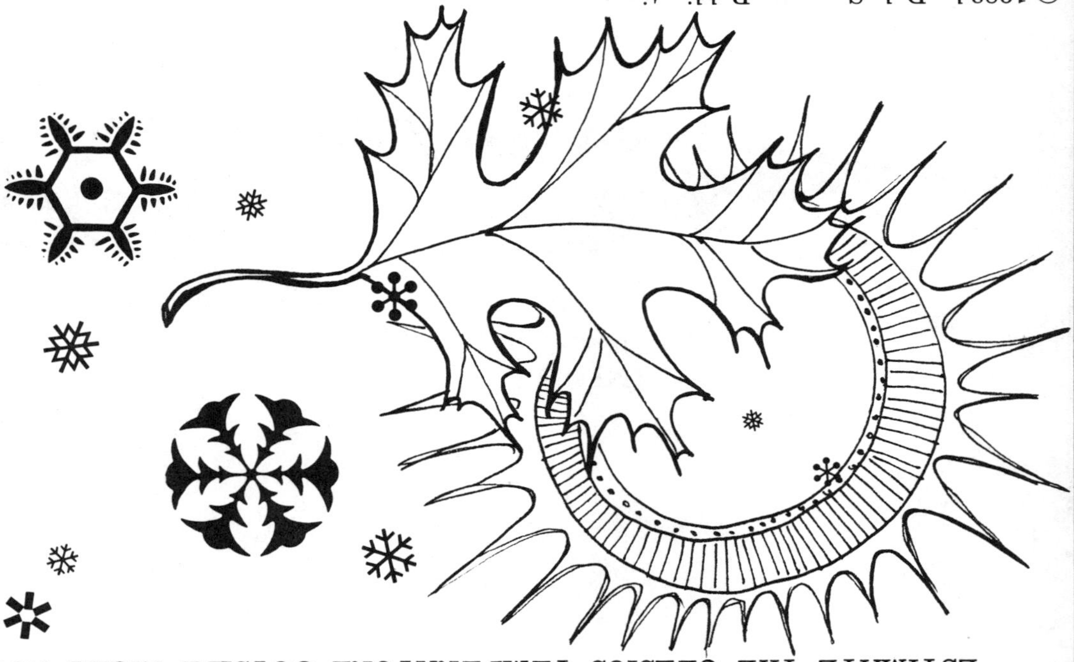

ESTIMATE THE CELSIUS TEMPERATURE OUTSIDE RIGHT NOW.

GREAT ESTIMATIONS 5E

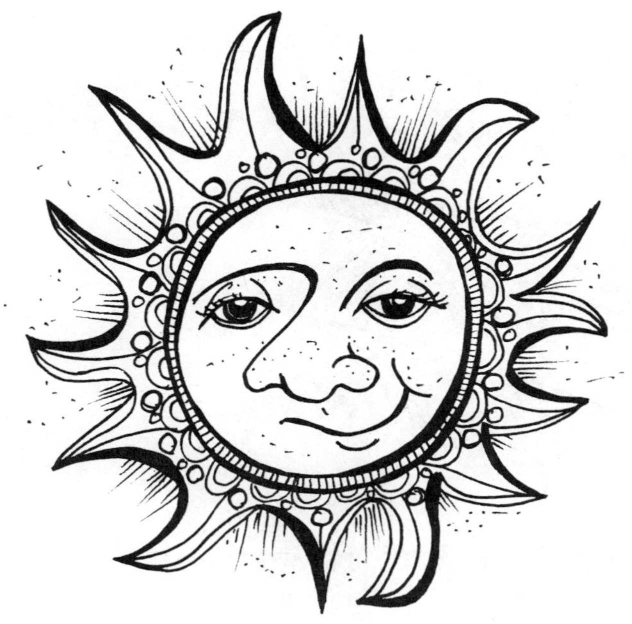

If it were very hot outside, what might the Celsius temperature be?

GREAT ESTIMATIONS 6E

If the temperature were $30^{\circ}C$ in this room,

HOW WOULD YOU FEEL?

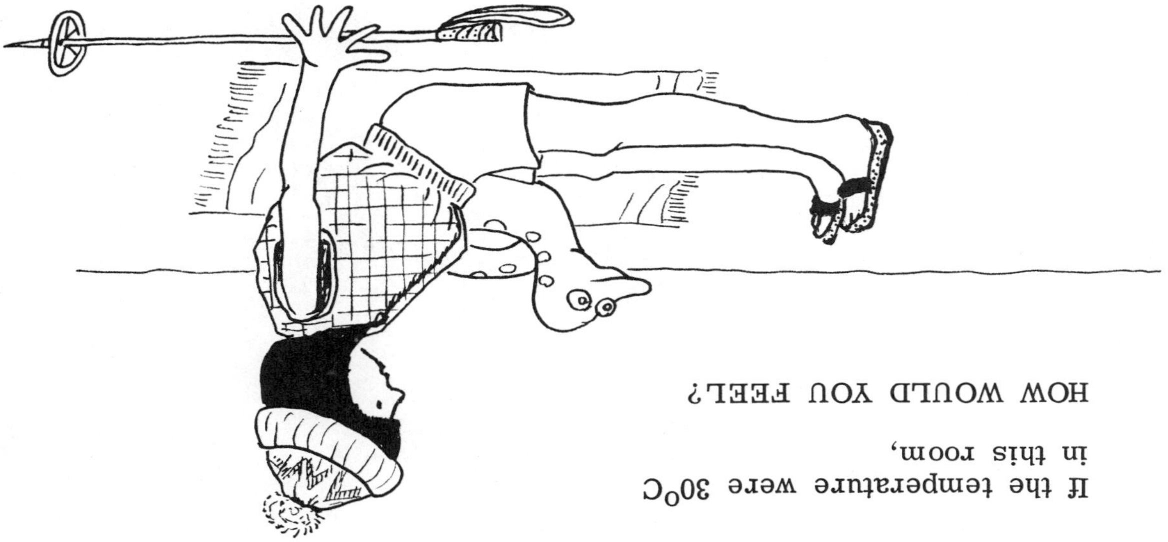

IF IT IS SNOWING OUTSIDE, WHAT MIGHT THE TEMPERATURE BE?

CARD
NUMBER

1. Circle - about 1/3 shaded
    Triangle - about 1/4 shaded
    Hexagon - about 3/4 shaded
    Trapezoid - about 7/8 shaded

2. Cloud - about 3/4 shaded
    Moon - about 1/3 shaded
    Star - about 4/5 shaded

3. The card measures about $21\frac{1}{2}$ cm by 14 cm, so an answer of about 300 sq. cm is reasonable.

4. Answers will vary.

5. Answers will vary.
    A dollar bill is about 6 inches by $2\frac{1}{2}$ inches.

6. About 3/4 of the shirt is dark.

Great Estimations

Which of these figures has about
1/3 of its area shaded,
1/4 of its area shaded,
3/4 of its area shaded,
7/8 of its area shaded?

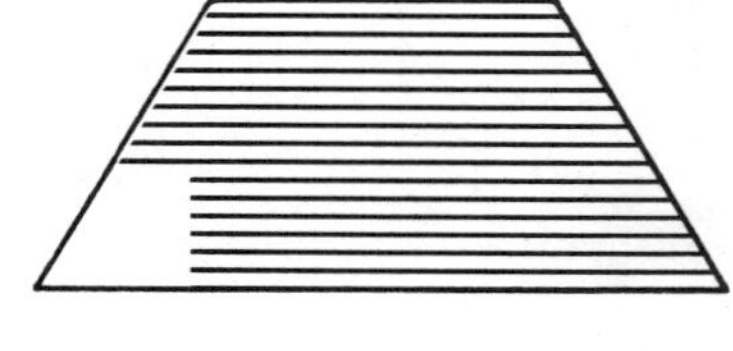

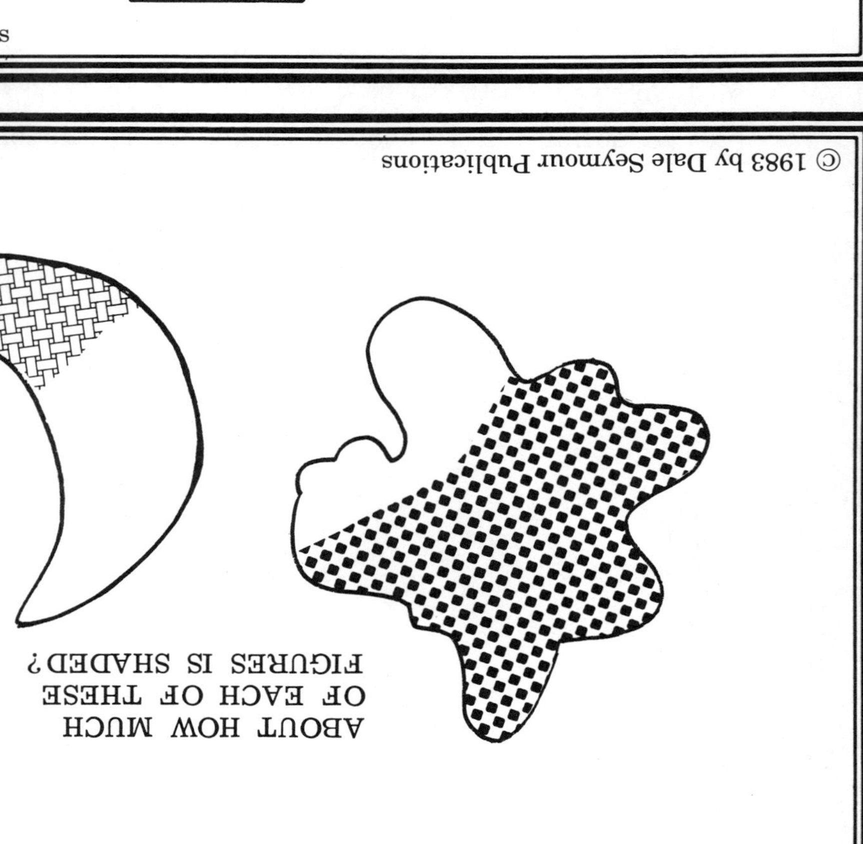

This is a square centimeter 

About how many square centimeters are there in this card?

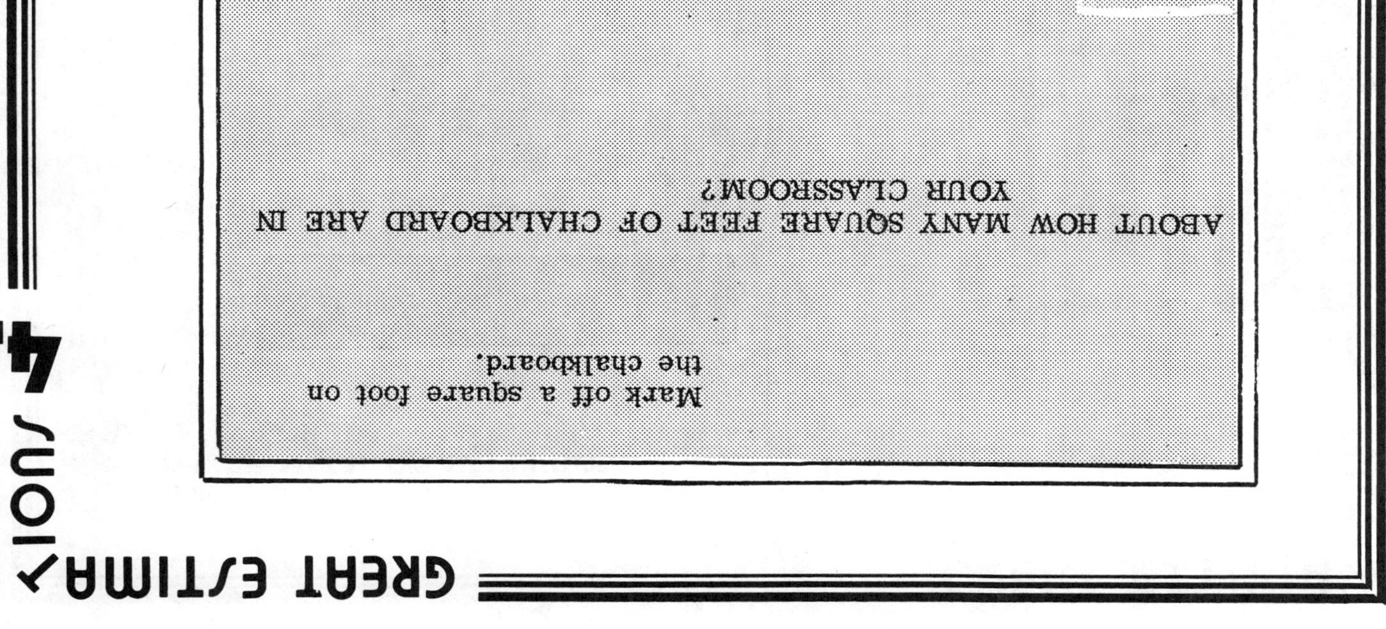

ABOUT HOW MANY SQUARE FEET OF CHALKBOARD ARE IN YOUR CLASSROOM?

Mark off a square foot on the chalkboard.

ABOUT HOW MANY DOLLAR BILLS
WOULD YOU NEED
TO COVER YOUR DESK?

Elroy bought a new shirt for school.
ABOUT HOW MUCH OF HIS SHIRT
IS DARK COLORED?

Hint: Look at the sample of
the shirt pattern.

Use this to estimate
the percent of dark
color in the whole shirt.

CARD
NUMBER

1. Students could do this in pairs or small groups if you don't want an entire class activity.

2. Prime time network television can not exceed 9 minutes 30 seconds of commercial time for every 60 minutes of programming. Non-prime time is different.

3. Read carefully.

4. Do this in pairs or small groups.

5. Have students make an estimate and then let them time each other.

GREAT ESTIMATIONS 1G

A CLASSROOM ACTIVITY

Students should close their eyes.
Say "START".
After X seconds, say "STOP".

Students should then estimate the time interval.

Let X vary from a few seconds to 1 or 2 minutes.

GREAT ESTIMATIONS 2G

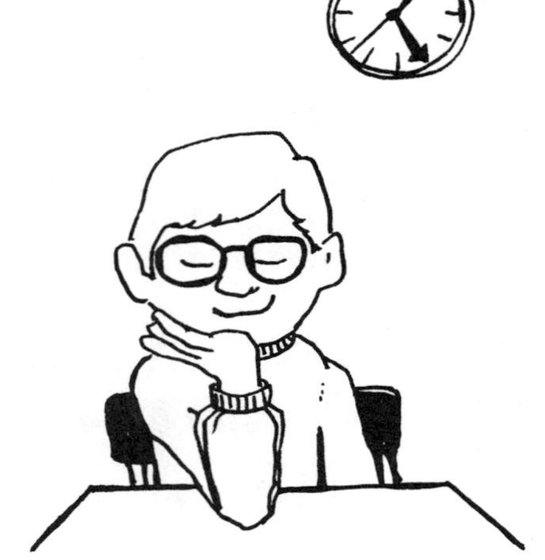

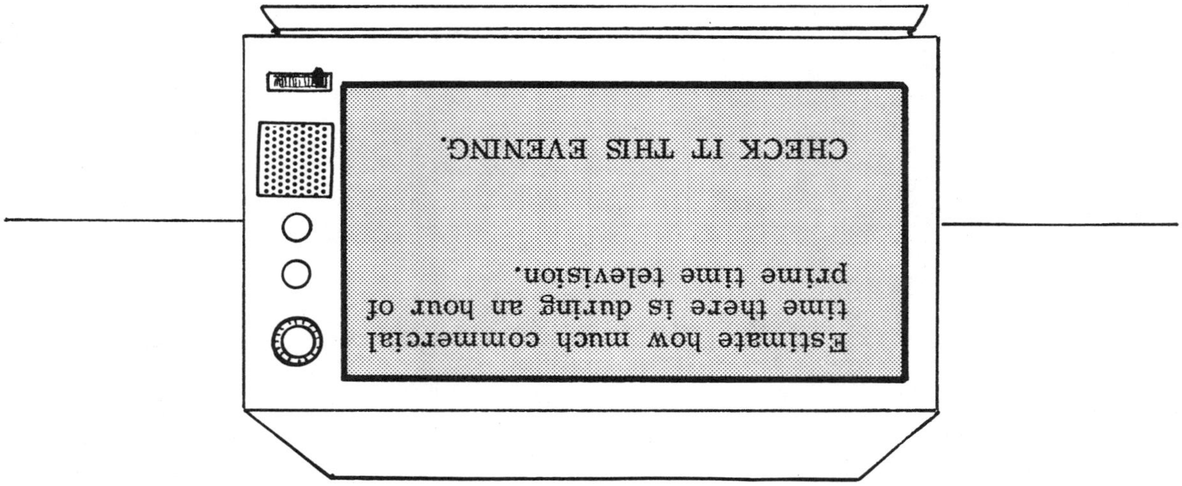

GREAT ESTIMATIONS 3G

1. Estimate your life expectancy, in years.
2. Now crumple a sheet of paper and throw it across the room.
3. If the teacher tells you to pick it up, say, "No - you pick it up!"
4. Now estimate your life expectancy, in seconds.

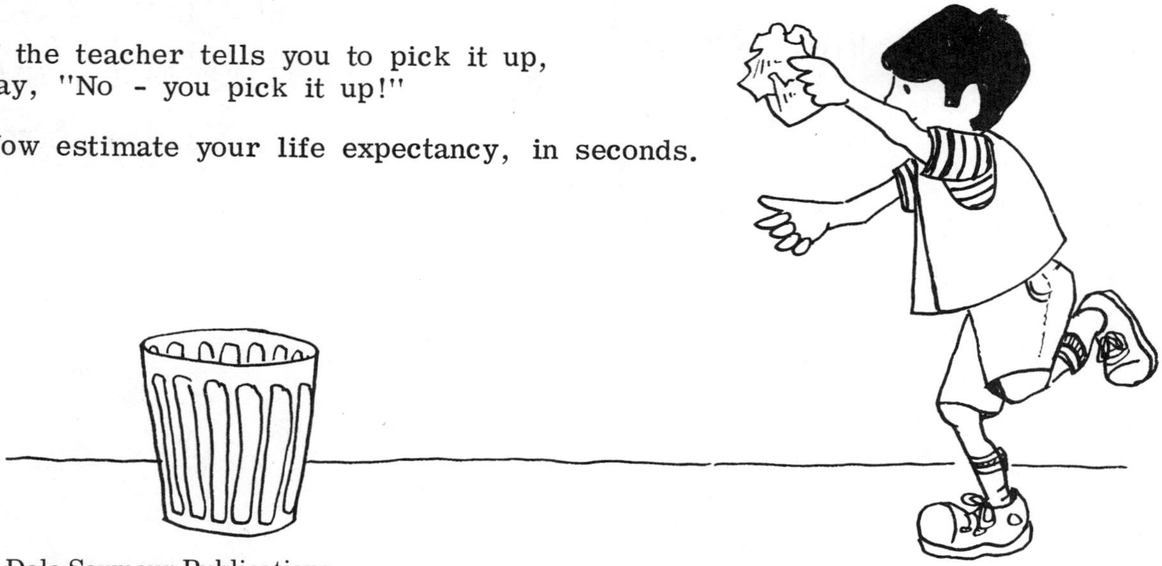

GREAT ESTIMATIONS 4G

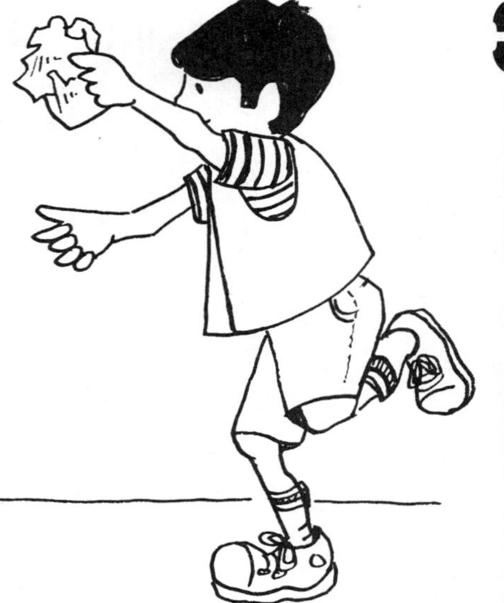

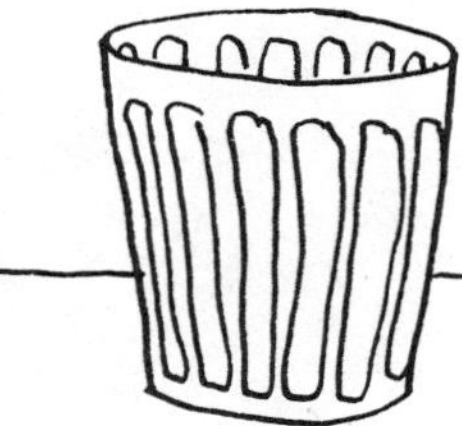

Estimate how long it will take to walk from the classroom door to the end of the hall and back.

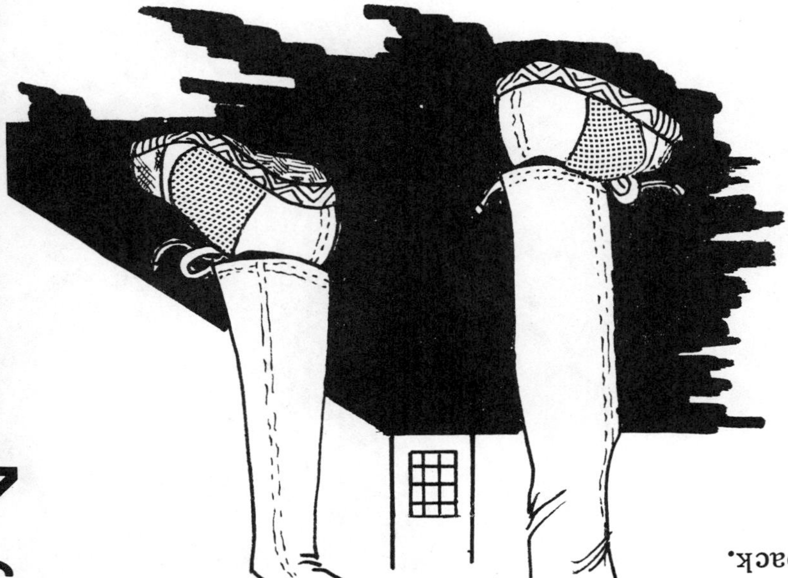

CHECK IT.

ABOUT HOW LONG DO YOU
THINK IT TAKES TO UNLOCK
THE COMBINATION LOCK ON
YOUR LOCKER?

(TIME IT!)